Multi-sensor Fusion for Autonomous Driving

Xinyu Zhang • Jun Li • Zhiwei Li • Huaping Liu •
Mo Zhou • Li Wang • Zhenhong Zou

Multi-sensor Fusion for Autonomous Driving

Springer

Xinyu Zhang
The School of Vehicle and Mobility
Tsinghua University
Beijing, China

Jun Li
School of Vehicle and Mobility
Tsinghua University
Beijing, China

Zhiwei Li
College of Information Science
and Technology
Beijing University of Chemical Technology
Beijing, China

Huaping Liu
Department of Computer Science
and Technology
Tsinghua University
Beijing, China

Mo Zhou
School of Vehicle and Mobility
Tsinghua University
Beijing, China

Li Wang
School of Vehicle and Mobility
Tsinghua University
Beijing, China

Zhenhong Zou
The School of Vehicle and Mobility
Tsinghua University
Beijing, China

ISBN 978-981-99-3279-5 ISBN 978-981-99-3280-1 (eBook)
https://doi.org/10.1007/978-981-99-3280-1

This Springer imprint is published by the registered company Springer Nature Singapore Pte Ltd.
The registered company address is: 152 Beach Road, #21-01/04 Gateway East, Singapore 189721, Singapore

Paper in this product is recyclable.

Foreword

In recent years, the popularity of intelligent connected vehicles has continued to soar, and has entered a period of rapid development worldwide. The expected intelligent of vehicles is based on existing vehicles and brings good automation based on connection methods through a new generation of mobile Internet technology. Such a development trend is still in the ascendant internationally, not only in the traditional automobile industry, but with the addition of Internet, communications, big data, and other enterprises to form a cross-border integrated corporate ecosystem.

From a technical point of view, the basic principles of realizing autonomous driving can be divided into three aspects: "sensors, data fusion, and 100% safe decision-making." Intelligent perception is the prerequisite for intelligent decision-making, and these two modules are inseparable from each other. As the pivotal foundation of autonomous driving, intelligent environment perception is the guarantee of driving safety and intelligence. The advanced three-dimensional environment perception system can promptly detect external things such as vehicles, pedestrians, obstacles, and roads that affect driving safety, and accurately obtain their information including three-dimensional position, size, driving direction, geometric shape, category, etc., to provide a basis for subsequent decision-making and control. Since it is difficult for a single detection means or sensor to robustly perceive complex scenes, the complementary advantages of multiple sensors can obtain more comprehensive and compatible driving environment information, thereby meeting the reliability and accuracy requirements of automated driving systems.

In the case of using multiple sensing devices, information fusion is a prerequisite to ensure driving safety. Environment perception technology based on multi-sensor fusion can improve the redundancy and fault tolerance of the system, leading to guarantee the rapid and correct information decision-making. From an industrial point of view, fusion perception technology is also a very critical technology in the field of autonomous vehicles. Previously, "Mims Consulting" reported that the demand for sensor fusion systems is expected to grow at a compound annual growth rate (CAGR) of approximately 19.4% in the next 5 years, and the market size is expected to reach 7.58 billion US dollars in 2023.

Recently, with the rapid development of artificial intelligence especially machine learning techniques, the area of fusion-based perception has revealed possibilities of its potential. To the best of my knowledge, this is the first book for multi-sensor fusion perception for autonomous driving. This work provides a good way of solving, in particular, the perception of autonomous driving under extreme conditions.

This book provides readers with an intuitive understanding and exciting applications of multi-sensor fusion perception, which is expected to play a key role in autonomous driving. I believe this book will reveal enormous practical impact as well as scientific insights into the research and education of multi-data fusion in autonomous driving.

Tsinghua University, Beijing, China
July, 2022

Keqiang Li

Preface

Autonomous driving research becomes a hot spot in the continuous development of the automotive industry, and accurate acquisition of environmental information is a vital foundation in technology applications. With the development of digitization and intelligence, the multi-modal fusion of different types of sensor data is an inevitable trend to realize all-weather and all-scenario autonomous driving perception. Multi-sensor fusion is the use of computer technology to automatically analyze and synthesize information and data from multiple sources under certain criteria to complete the information processing procedure required for decision-making and estimation.

In this process, it is required to make full use of multi-source data for reasonable control and use, while the ultimate goal of information fusion is to derive more useful information based on the separated observation information obtained by each sensor, through the multi-level and multi-view combination of information. This not only takes advantage of the cooperative operation of multiple sensors, but also comprehensively processes data from other information sources to improve the intelligence of the entire sensor system. Therefore, the research of multi-modal data fusion method deserves important academic and application value.

Due to the different ways of recording information, there is often complementarity between different sensors, which can be used to improve the robustness of the perception model. Taking working conditions with poor lighting conditions as an example, it is difficult for RGB cameras to record effective information, while active sensing sensors, such as Light Detection and Ranging (LiDAR), Radar, and Depth Cameras, are not easily affected by external environmental conditions. For the sparse point cloud of LiDAR and Radar, the obtained low-resolution data is difficult to be used for high-precision detection, while RGB images can provide dense data. Consequently, understanding and using the relationship and complementarity between multi-modal data has become a critical issue for multi-modal data fusion.

Data fusion in autonomous driving can play a role in multiple tasks, such as object detection, semantic segmentation, object tracking, and simultaneous localization and mapping. At present, the most common fusion mode in high-level autonomous driving is the fusion of LiDAR and RGB camera. The LiDAR point cloud can actively perceive objects in a larger range without being affected by the lighting conditions, while the color, texture, and other visual information provided by the RGB image can be used for higher-precision visual tasks.

While the perception method based on multi-sensor fusion brings benefits to the autonomous driving system, it also derives a variety of problems and challenges, such as the calibration between multiple sensors, the response to sensor failures, and the exploration of more fusion methods. On the other hand, the existing fusion methods lack a unified and clear definition, and various algorithms still have differences in specific implementation details, resulting in the effect of the fusion step on the model that cannot be well reflected in the integration model. In general, the current progress has not yet clarified the promotion of multi-modal data fusion perception for autonomous driving technology, which requires in-depth discussion and analysis to support the development of autonomous driving technology.

To meet the aforementioned challenges of autonomous driving fusion perception, the main body of the book is divided into three parts: Basic, Method, and Advance. The first part, Basic, Chaps. 1 and 2 start from the mechanism of data fusion, comprehensively review the development of automatic perception technology and data fusion technology, and give a comprehensive overview of various perception tasks based on multimodal data fusion. In the second part, Method, for various autonomous driving perception tasks, a series of innovative algorithms are proposed in Chaps. 3–6 to effectively improve the accuracy and robustness of autonomous driving-related tasks, and provide ideas for solving the challenges in multi-sensor fusion methods. Furthermore, in the third part, Advance, to transition from technical research to intelligent connected collaboration applications, Chaps. 7–9 propose a series of exploratory contents such as practical fusion datasets, vehicle-road collaboration, and fusion mechanisms. The last Chap. 10 summarizes this book and presents some prospects.

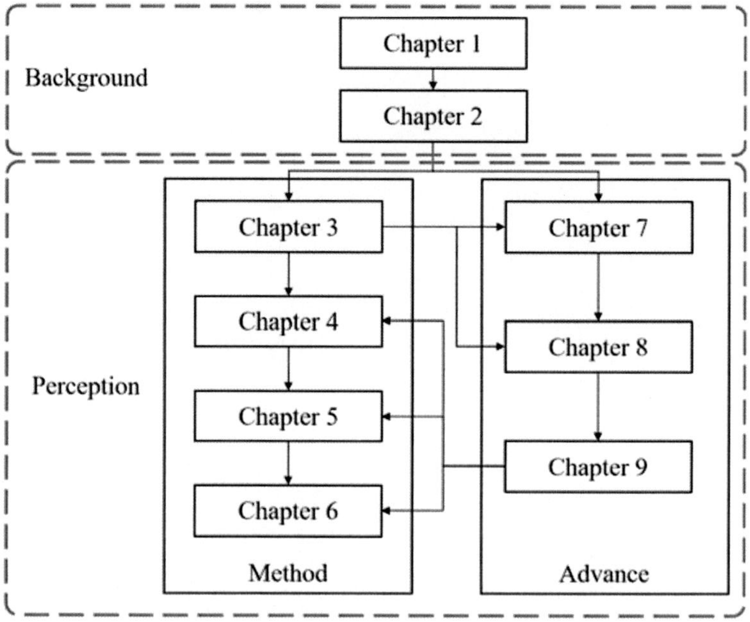

Fig. 1 Organization of the book: logical dependency among parts and chapters

This book is suitable as a reference book for graduate students with a basic knowledge of machine learning as well as professional researchers interested in fusion perception and autonomous driving, and machine learning (Fig. 1).

Beijing, China
July, 2022

Xinyu Zhang
Huaping Liu
Jun Li

Acknowledgments

This book refers to our research work at State Key Laboratory of Automotive Safety and Energy, Tsinghua University, Beijing, China, and State Key Laboratory of Intelligent Technology and Systems, TNLIST, China.

Five years ago, we started looking into the challenging field of multi-sensor fusion perception. Starting from the multimodal perception research of Dr. Jianhui Zhao, post-doctoral fellows and researchers in the research group including Dr. Qifan Tan, and Shichun Guo have joined the field of fusion perception research. By designing various innovative fusion algorithms, we have gradually exploited the advantages of multi-sensor fusion in autonomous perception systems. We would like to thank everyone who have participated for their support, dedication, and cooperation.

We would like to sincerely thank our interns Baowei Xu and Ganglin Tian for successfully applied multi-sensor information fusion to the detection of three-dimensional targets and tracking, and the excellent results formed provided important support for our research. In addition, we also thank our interns Shifan Zhu, Wenju Gao, and Yijin Xiong for their valuable help in multi-sensor calibration and fusion localization, which is a great inspiration to the authors.

We would like to express our sincere gratitude to Yizhe Li, Dan Xia, and Haobo Yang, who have provided immense help with preparation of figures and with the proofreading of the book.

The completion of this book cannot be separated from the strong support of Springer Publishing. We would like to thank our commissioning editors, Lanlan Chang and Veena Perumal, for their great supports.

A great deal of this research was supported by the National High Technology Research and Development Program of China under Grant No. 2018YFE0204300 and the National Natural Science Foundation of China under Grant No. 62273198.

Tsinghua University, Beijing, China
July, 2022

Xinyu Zhang
Huaping Liu
Jun Li

Contents

Part I
Basic

This part of the book comprises two chapters. In Chap. 1, the basic concepts of autonomous driving and environment perception are presented. The historical development of fusion perception technology and common datasets are also introduced. Chapter 2 serves as a basis of the whole book, by providing an overall review of multimodal fusion perception technologies in autonomous driving.

Chapter 1
Introduction

Abstract Autonomous driving is the product of the deep integration of the automobile industry and the new generation of information technology. Among them, environmental perception technology is one of the vital essential technologies for realizing automatic driving. It is the premise of realizing safe and intelligent driving in complex road conditions. Existing environmental perception technologies for autonomous vehicles are generally based on sensors such as vision sensors, millimeter-wave RaDARs, and LiDARs. This chapter aims to gain an understanding and mastery of autonomous driving perception and critical developments.

1.1 Autonomous Driving

Autonomous driving refers to vehicles equipped with advanced intelligent systems and various sensing devices (including cameras, RaDARs, navigation devices, etc.) enabling them to have functions such as complex environmental perception, intelligent decision-making, collaborative control, and execution. The autonomous driving system can achieve safe, energy-saving, and efficient driving and can eventually replace human operators.

The classification scheme proposed by the Society of Automotive Engineers (SAE) J3016 document is currently a generally accepted standard, which divides autonomous driving technology into six levels from L0 to L5. The system at the L0–L2 level is defined as a "driver assistance system." Regardless of whether the driver assistance feature is on or not, the driver is in complete control of the vehicle and is completely responsible for its safety with this degree of support. Common AEB active braking, lane departure warning, lane-keeping, ACC adaptive cruise, and other functions belong to this level of functions. The L3–L5 level system is usually defined as an "autonomous driving system." When the system is activated at this level, the control of the vehicle is dominated by the vehicle.

In other words, autonomous driving entails that at least certain safety-critical control operations (such as steering, acceleration, and braking) can be accomplished automatically without the driver's direct input. Autonomous driving vehicles generally use onboard sensors, GPS, and other communication technology equipment to

X. Zhang et al., *Multi-sensor Fusion for Autonomous Driving*,
https://doi.org/10.1007/978-981-99-3280-1_1

obtain information, make decisions and plans for safety conditions, and implement control appropriately to a certain extent.

The autonomous driving system is a comprehensive system that integrates advanced artificial intelligence and information control technology, combined with functions such as environmental perception, driving cognition, planning control, and advanced driving assistance systems. In response to driving requirements in multiple scenarios, the autonomous driving system has increasingly strict requirements for environmental perception. In the architecture of the autonomous driving system, the sensor layer is compared to the "eyes" of the vehicle, which includes the camera-based vision sensor and RaDAR sensors such as millimeter-wave RaDAR, LiDAR, and ultrasonic RaDAR. The localization and attitude estimation of the vehicle body mainly rely on positioning sensors such as the Global Positioning System (GPS), BeiDou Navigation Satellite System (BDS), and inertial navigation system (INS). These sensors can obtain information about the vehicle, including driving velocity and attitude, and provide effective data for the positioning and navigation of the autonomous vehicle.

Subsequently, based on the comprehensive processing of the data sensed by the sensor layer, autonomous decision-making and path planning tasks are performed, and the control instructions are passed to the bottom-level actuators to complete the entire autonomous driving process. The architecture of the autonomous driving system is shown in Fig. 1.1.

Although each research unit has different divisions of the autonomous driving system architecture and different focuses in technical research, it covers seven aspects: sensors, perception, decision-making, control, human-computer interaction, public services, and execution.

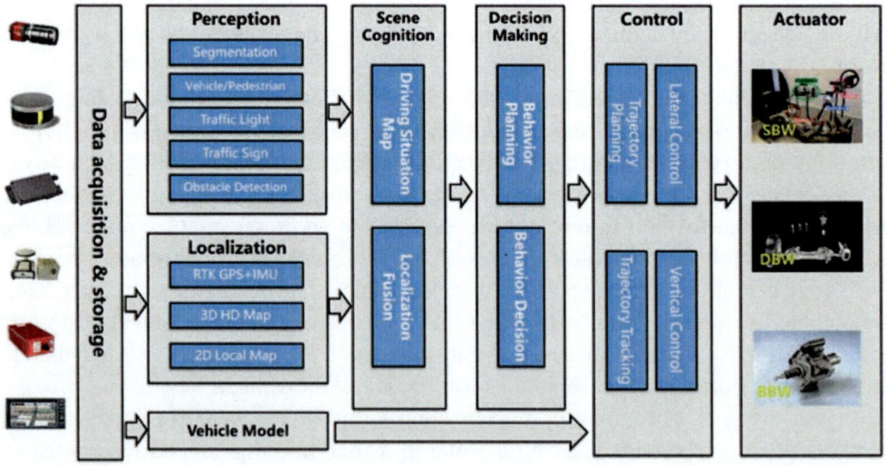

Fig. 1.1 An architecture of the autonomous driving system

Sensors: It is composed of RaDAR sensors, vision sensors, GPS, body sensors, etc., for collecting essential driving data. To realize the plug and play of these sensors, it is necessary to standardize the data format of various sensors. That is, to convert the unique data format of the sensor into a standard format that can be processed by the smart car. This layer sends the collected sensor data to the perception module for further processing.

Perception: It performs the task of analyzing sensor data and realizing specific functions such as object detection, road segmentation, and body state estimation. As the first data processing module, it provides basic support for the planning and decision-making module.

Decision-making: It mainly completes path planning and navigation. By analyzing the environmental data obtained from the perception module, the driving mode of the autonomous driving vehicle is determined. It determines the location of the vehicle on the fine electronic map and generates a driving trajectory based on the coordinates of the target point. At the same time, the influence of both human intervention and obstacle conditions on trajectory generation requires to be considered.

Control: It controls the vehicle to follow the trajectory based on the trajectory data and the current vehicle status. In the meantime, it receives human intervention instructions to perform acceleration, deceleration, and steering operations. This layer directly outputs control commands to the accelerator, brake, and steering controllers of the vehicle.

Human-computer interaction: It receives touch commands and emergency braking commands from the driver and outputs them to the control layer. Simultaneously, the environment and vehicle information will also be fed back through sound and images as a reference for the driver.

Public services: It provides services for the above layers, including data communication, data recording, map file reading, and writing. Execution: It is directly associated with the electronic control module of the vehicle and executes driving actions based on the received control commands, such as adding or subtracting the throttle, electric steering operation, and power control.

1.2 Sensors

In the autonomous driving system, the sensor hardware devices used for environmental perception are diverse. Currently, the majority of sensor configuration schemes utilized in autonomous driving vehicles are frequently a combination of numerous models or equipment types. Generally, as the number of sensors rises, sensing precision and range synchronization improve, but costs rise. The sensors of autonomous driving vehicles can be divided into vision sensors, RaDAR sensors, positioning sensors, auditory sensors, and attitude sensors. For the environment perception task of autonomous driving, the most commonly used sensors are cameras, LiDARs, and millimeter-wave RaDARs.

The visual sensor primarily involves the use of a camera to detect, track, and identify targets (vehicles, pedestrians, traffic signs). The specific function can be disassembled to perceive obstacles and drivable areas around the vehicle, understand the semantics of traffic signs and road markings, and understand the current driving scene. Compared with other sensors, industrial cameras required for computer vision are relatively mature in terms of technology. From the perspective of hardware, cameras have the characteristics of high stability, strong transmission ability, and strong anti-interference ability. Due to its price advantage, multiple cameras are widely used in the ADAS market to meet driving needs.

As an active imaging RaDAR technology developed based on laser rangefinders, Light Detection and Ranging (LiDAR) emits and receives laser beams and analyzes the return time of the laser after encountering the target object to calculate the relative distance. In the meantime, based on the surface information of the target object collected in this process, various related data including the three-dimensional coordinates, reflectivity and texture, and the three-dimensional model can be quickly obtained. Therefore, a three-dimensional point cloud map of the entire environment can be established to achieve the purpose of environmental perception. Given the potentially short time of flight due to the speed of light, the measurement equipment is required to have very high accuracy. From the perspective of effects, the more the laser RaDAR beams (dimensions), the higher the measurement accuracy, and the higher the safety.

The essence of the vehicle-mounted millimeter-wave RaDAR is a frequency-modulated continuous-wave ranging RaDAR. With the characteristics of simple structure and small size, this type of RaDAR can obtain the relative distance and relative speed of the measured target at the same time. The basic principle of RaDAR is that when the transmitted continuous-frequency modulation signal encounters the target, it will generate an echo with a certain delay from the transmitted signal. After that, the frequency mixing process is performed by the mixer, and the result after mixing is related to the relative distance and relative speed of the measured target. The frequency bands of vehicle-mounted millimeter-wave RaDARs are mainly concentrated in the two frequency bands of 24GHz and 77GHz. For instance, the ESR millimeter-wave RaDAR developed by Delphi Corporation of the United States has the functions of mid-range scanning and long-range scanning at the same time. Adopting the continuous modulation method and applying the Doppler test principle, it can obtain the relative distance, angle, and speed of 64 targets within the farthest range of 174 m.

1.3 Perception

The main research direction of perception technology in autonomous driving is real-time perception and understanding of the surrounding environment. This technology still faces the challenge of processing large amounts of data from multiple sensors, such as data from cameras, wireless communication devices, ranging RaDARs, and infrared devices. The data of autonomous driving vehicles is usually collected by multiple sensors and preprocessed to form various features to detect and track static and dynamic targets in the environment. In addition, some inferences can be performed, such as vehicle behavior and scene understanding. The main function of environment perception is to realize the detection and tracking of lanes, roads, and traffic participants based on the onboard hardware system.

As one of the tasks in the field of computer vision, object detection is the basis of many other vision tasks, such as instance segmentation [8] and target tracking [10], aiming to detect each instance of different types of objects. Considering the availability of data and the richness of data features, it generally refers to target detection with RGB images as the main data. Typically, the detection result is expressed as the bounding boxes labeled on images to locate objects and the probability of object category properties [20]. According to the definition, it is usually required that the object detection algorithm first search for the area that may contain the target on the image and then classify the area, which is a multi-stage model. With the development of deep learning, a one-stage detection model is proposed, which can perform classification at the same time as detection, thereby increasing the detection speed. In addition, the detection methods are also inconsistent for different modal data. As multiple sensors such as LiDAR, RaDAR, and RGB-depth camera are applied to autonomous vehicles, it is also necessary to pay attention to methods based on point clouds provided by RaDAR, depth images, or other modal data in autonomous driving. For object detection on images, the accuracy of traditional methods is often inferior to deep learning methods, while the latter often requires large datasets and long-term training to learn features. For object detection on the point cloud, the advantage is that it can use three-dimensional spatial information for detection. However, the disadvantage is that the increase in spatial dimensions causes the point cloud data to be too sparse, resulting in poor model fitting effects [1]. The depth image, that is, the RGB image with distance information recorded by the depth camera [6, 15], combines the characteristics of the image and the point cloud, but it has not yet become the mainstream due to the insufficient performance of the camera. In summary, although object detection methods for multiple modal data are constantly being proposed, most of the models are still based on images. The current model mainly evaluates the effect by the coincidence of bounding boxes [17] and determines whether the prediction is correct by setting a threshold, namely, intersection over union (IoU).

After object recognition, object tracking is a crucial processing step that tries to monitor the target of interest and generate its trajectory. According to the generation method of the motion trajectory, the object tracking algorithm can be divided into offline algorithm and online algorithm [14]. The offline tracking algorithm can use future frames to optimize the tracking results using global information. Nevertheless, the autonomous driving system itself requires real-time data processing; therefore, online monitoring techniques are typically employed. The online tracking algorithm accepts the sensor data and detection results of the current frame and historical frames and associates the detection results with the existing historical target trajectory [16]. In recent years, tracking methods based on correlation filtering and tracking methods based on deep learning have developed rapidly. Compared with traditional optical flow method, Kalman filter method, mean shift method, and other traditional algorithms, correlation filtering algorithm tracking has advantages in speed, while deep learning methods have higher accuracy [19]. The object tracking based on correlation filtering [5] can effectively implement correlation calculations through fast Fourier transform, and its calculation speed can often reach hundreds of frames per second. Due to the fact that this method typically relies on lower-level features, such as color and texture, its reliability and robustness are far inferior to those of methods based on deep learning. The tracking method based on deep learning has achieved better tracking effect than traditional methods, by using deep neural network to extract the deep features of the image. With the development of computer performance in recent years, object tracking methods based on deep learning can meet the real-time requirements of autonomous driving. Nevertheless, due to the peculiarities of autonomous driving scenarios, such as mutual occlusion and interaction between traffic participants, as well as frequent changes in the driving environment, single-modal data is frequently flawed, resulting in insufficient tracking trajectories. Therefore, some researchers overcome the problem of single-sensor data defects by considering the redundancy and complementarity of multi-sensor data. Different from object detection, the object tracking multi-sensor fusion algorithm can not only use different sensor data for multi-sensor fusion but also fuse the historical information of the sensor itself or cross-fuse the historical information of different sensors.

1.4 Multi-Sensor Fusion

Table 1.1 summarizes the types, functions, advantages, and disadvantages of several common sensors configured on autonomous driving vehicles.

Regardless of whether it is a monocular camera, a binocular camera, a multi-eye camera, or a depth camera, regardless of the number of pixels or sampling rate, it cannot solve all the problems of image processing. Due to the diversity and complexity of the road environment and weather, as well as the motion characteristics of autonomous driving vehicles, the camera is susceptible to many uncertain factors such as illumination, angle of view, scale, shadow, fouling, and background interference and target occlusion. In the process of autonomous driving, traffic elements such as lane lines and signal lights are worn to a certain degree, and reflections are normal, so there is no perfect camera.

Radar sensors have strong robustness to interference factors such as light and color. LiDAR, millimeter-wave RaDAR, and ultrasonic RaDAR also have their advantages. However, the number/type of RaDAR installed and the high sampling frequency cannot completely solve the detection problems under severe weather conditions such as pit reflection, smoke and dust interference, rain, snow, and fog. It is also difficult to achieve true all-weather, all-time, and all-road conditions, which means that the RaDAR sensor is also imperfect.

Table 1.1 Comparison of common sensors configured on autonomous driving vehicles

Sensor type	Detection distance (m)	Pedestrian detection	Object detection	Object recognition	Function	Advantages	Disadvantages
Camera	50	✓	✓	✓	Rely on machine vision algorithms to detect and recognize the surrounding environment and objects, detect the distance of obstacles in front, positioning and environmental modeling	Able to identify the type of object and the color of the signal light; low price cost	Affected by light conditions; depends on learning samples
LiDAR	200	✓	✓	×	Static object detection, dynamic object detection and tracking, road surface detection, positioning and environmental modeling	Able to detect most objects, with high distance accuracy	Affected by heavy rain, heavy snow, and other severe weather
Millimeter-wave RaDAR	170	×	✓	×	Motion detection of most vehicles, mostly used for adaptive cruise and collision warning	Less affected by rain, snow, and fog; lower cost	Poor detection of pedestrians

Different types of sensors have their own advantages and disadvantages, and neither the “millimeter-wave RaDAR + camera” solution nor the “3D LiDAR” solution could be unique. The various configuration schemes seem to be competing, but they are complementary, inspiring, and supplementing each other. For different sensing tasks, different sensor types and models are required. To fulfill the driving task, it is not essential to configure all of the most expensive and many sensors. Sensor configuration should be task-oriented, that is, targeted selection and combination to achieve the most cost-effective optimal configuration. At the same time, for autonomous vehicles to achieve safe driving, multi-sensor collaboration and information redundancy must be guaranteed. Therefore, multiple sensors are often required to work collaboratively and complement each other’s advantages. Various sensors cooperate with each other to input the collected environmental data in the form of pixels and point clouds. Furthermore, artificial intelligence algorithms are used for extraction, processing, and fusion, to further form a complete driving situation around the vehicle, and provide a basis for the behavioral decision-making of autonomous driving vehicles.

For object detection tasks, according to Di Feng et al. [7], data fusion occurs at different stages of the object detection model; thus, the fusion method can be divided into early fusion, middle fusion or deep fusion, and late fusion. Specifically, early fusion focuses on the fusion of original data or data that has only been preprocessed, and late fusion fuses the calculation results of multiple branches of the model to obtain the final result. Combining the characteristics of early fusion and late fusion, middle fusion combines the data of multiple modalities or their corresponding feature maps for continued analysis. Middle fusion can not only exist in the intermediate steps of the model but can also run through the entire model. The three types of fusion are also known as data-level fusion, feature-level fusion, and decision-level fusion in other publications [13]. However, in the deep learning model with multi-level and multi-model combination, the result of model branching is also the intermediate feature of the whole model [3]. Most fusion methods are classified as middle fusion, which leads to the lack of discrimination in the division method by stage. To distinguish the degree of model fusion more intuitively, a classification scheme for fusion methods is produced from the perspective of fusion levels, namely, data fusion, feature fusion, result fusion, and auxiliary estimation, as shown in Fig. 1.2. According to the proposed classification method, the division of the fusion method is no longer based on the sequence of the fusion steps in the model, but the role of the fusion result in the model. Moreover, the division of fusion will not be restricted by the form of the fusion data, which more intuitively reflects the effect of the fusion operation on the model.

Similarly, for object tracking tasks, multi-sensor fusion mainly occurs at three levels, data-level fusion, feature-level fusion, and decision-level fusion, as shown in Fig. 1.3. Zhang et al. [18] considered the RGB and TIR modalities and used the DiMP tracker [4] as a benchmark to quantitatively analyze and verify the effect of three different fusion levels on the performance of the fusion tracking algorithm. Linder et al. [11] fused HOG features and depth features in four different tracking algorithms [2, 9, 12] and proved that these four tracking algorithms have different

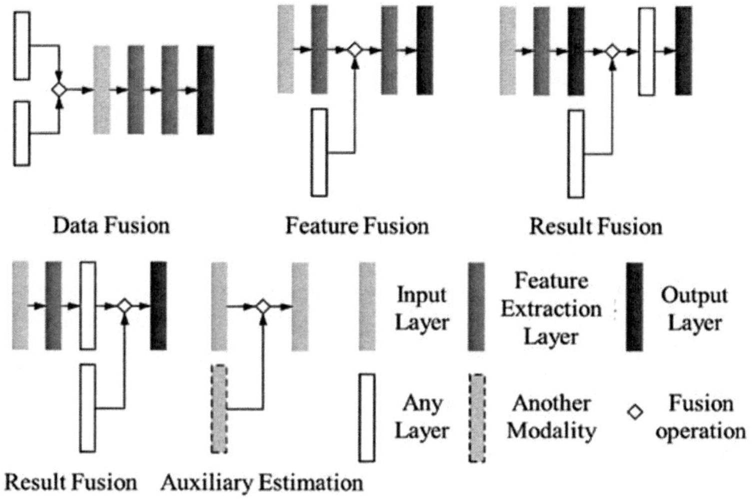

Fig. 1.2 Examples of different levels of fusion for object detection

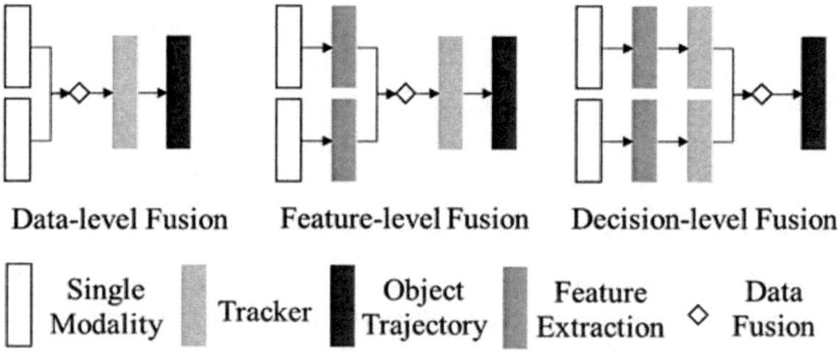

Fig. 1.3 Examples of different levels of fusion for object tracking

degrees of improvement. The pervious experiments all indicate that the performance of tracking algorithms based on sensor fusion has been improved and, compared to data-level fusion and decision-level fusion, feature-level fusion improves tracking performance more significantly.

1.5 Public Datasets

In recent years, deep learning has made great progress in scene understanding. Deep learning can automatically learn object features based on labeled data. The more labeled data used for training and the more comprehensive the features of various types of objects are included, the more sufficient the features that the deep learning network can learn. For complex object features, deep learning can be fitted with huge parameter models, while it is often difficult for people to display and define complex target features completely and accurately.

High-quality annotation data is the fuel that drives deep learning algorithms, and there are currently a large number of open-source datasets based on unmanned platforms:

(1) The KITTI dataset was co-founded by the Karlsruhe Institute of Technology in Germany and the Toyota Technological Institute at Chicago, which is currently one of the largest computer vision algorithm evaluation datasets in the autonomous driving scenario in the world. This dataset contains rich and diverse sensor data (such as binocular cameras, 64-line LiDAR, etc.) and a large number of calibration true values (including detection of 2D and 3D label frames and tracking of trajectories). To cover more real driving conditions, the collected scenes mainly include urban areas, villages, and highways. Each image has a maximum of 15 cars and 30 pedestrians and various degrees of occlusion and interception. The entire dataset consists of 389 pairs of stereo images and optical flow diagrams, 39.2 km visual ranging sequence, and more than 200k 3D annotated object images, which are sampled and synchronized at a frequency of 10 Hz. The types of tasks that can be studied based on this dataset include stereo evaluation, optical flow evaluation, visual odometry, 3D object detection and tracking, lane detection, semantic segmentation, etc.
(2) Waymo, an autonomous driving company under Google's parent company Alphabet, has open-sourced its autonomous driving database named Waymo Open Dataset to help the research community achieve breakthroughs in machine perception and autonomous driving technology. The dataset covers dense urban and suburban environments in Phoenix, Arizona, Washington, California, etc. and captures data under various driving conditions (including day and night, dawn and dusk, and sunny and rainy days). The Waymo acquisition platform is equipped with 5 cameras and 5 LiDARs, which collected 3000 driving records with a total of 600,000 frames. Vehicles, pedestrians, cyclists, and signs have been carefully labeled, with approximately 12 million 3D labels and 1.2 million 2D labels, while 113k LiDAR object trajectories and 160k camera image trajectories have been generated. All truth boxes contain tracking identifiers that support object tracking, and such consecutive frames can facilitate researchers to develop models to track and predict the behavior of other road users. These data can help researchers make progress in the fields of scene adaptation, scene understanding, and behavior prediction.
(3) The nuScenes dataset is a large-scale autonomous driving dataset established by the autonomous driving company nuTonomy. It collects data from six cameras, one LiDAR, and five millimeter-wave RaDARs, which is the only dataset with RaDAR data that can be found at present. This dataset consists of 1000 scenes; each scene is 20 s in length, including high traffic density (such as intersections and construction sites), rare objects (such as ambulances and animals), and potentially dangerous traffic conditions (such as the chaotic crossing of roads),

maneuvering (such as changing lanes, turning, and stopping), etc. The nuScenes dataset contains 1.4 million images, 400,000 LiDAR scans, and 1.1 million three-dimensional bounding boxes.

(4) The Cityscapes dataset was jointly launched in 2015 by three German companies including Daimler, which is the urban landscape dataset, which provides image segmentation datasets in an unmanned driving environment. It is used to evaluate the performance of vision algorithms in the semantic understanding of urban scenes. The Cityscapes dataset contains 50 cities with different scenes, different backgrounds, and street scenes in different seasons, which provides 5000 finely labeled images, 20,000 roughly labeled images, and 30 types of labeled objects. This dataset is currently recognized as one of the most authoritative and professional image semantic segmentation evaluation sets in the field of autonomous driving. It focuses on the understanding of the urban road environment in real scenarios, and the task is more difficult and closer to popular requirements such as autonomous driving.

(5) In March 2018, Baidu's large-scale autonomous driving dataset ApolloScape was opened, and it is committed to providing more practical data resources and evaluation standards for autonomous driving technology researchers around the world. The acquisition platform is equipped with two LiDARs and six cameras. ApolloScape has opened 147,000 frames of pixel-level semantic annotation images, including hundreds of thousands of frames of pixel-level semantic segmentation and high-resolution image data for perceptual classification and road network data, as well as corresponding pixel-by-pixel semantic annotations, covering a 10KM area around 3 sites from 3 cities. Moreover, each area was repeatedly collected under different weather and lighting conditions. By adding more sensors to expand the diversity of data, it is committed to creating a platform with the highest degree of real-world reproduction and the most abundant scenes. ApolloScape is currently the 3D autopilot public dataset in the industry, with the most complex environment, the most accurate labeling, and the largest amount of data. In terms of data quality, the precision of data labels in ApolloScape exceeds that of the same type of KITTI and Cityscapes datasets.

(6) In 2020, AI labeling company Scale AI and Hesai Technology jointly released a dataset called PandaSet. Combining the advantages of hardware and annotations, it is a high-quality dataset for L5 autonomous driving. The sensor suite for data collection mainly includes one mechanical LiDAR, one solid-state LiDAR, five wide-angle cameras, and one telephoto camera, which fully captures the complex and changeable environmental factors in urban areas. The dataset includes more than 48,000 camera images and 16,000 LiDAR scanned point cloud images. The most challenging driving conditions in level 5 autonomous driving are covered, including all-day light conditions and weather conditions. In addition, the complex urban environment, dense traffic and pedestrians, buildings, greening, and other facilities are also considered in this dataset. Nowadays, autonomous driving datasets are still dominated by pictures and

LiDAR, but as more companies join, the public datasets will become more and more abundant.

As research investment in autonomous driving continues to grow, public datasets are expected to become more abundant. The development of autonomous driving datasets is currently focused on several key areas. Firstly, more scenarios and contexts can be added to improve the adaptability and generalization of models. Secondly, multiple types of datasets can be integrated to increase the quality and stability of data, facilitating the creation of comprehensive solutions and robust evaluation of autonomous driving tasks. Lastly, the promotion of dataset sharing and open-source approaches can help accelerate the development and diffusion of autonomous driving technologies.

1.6 Challenges

The Defense Advanced Research Projects Agency (DARPA) of the United States held the Grand Challenge in 2004 and 2005 and the Urban Challenge in 2007 to test and promote the development of intelligent driving technology. In the 2004 competition, participating vehicles were required to successfully traverse 200 miles of rugged and varied terrain within 10 h, but none of the teams ran the entire course. In 2005, the race was held in the Nevada desert in the Southwestern United States. It was consistent with the previous rules and required autonomous vehicles to complete the race's prescribed route within 10 h. The entire race route will not exceed 175 miles, with artificial obstacles added in addition to the natural terrain. The Stanley driverless car from Stanford University finally successfully traversed 132 miles of rugged desert with a best time of 9 h and 55 min and became the champion of this competition. Two participating teams from Carnegie Mellon University, Red Team and Red Team Too, ranked second and third with 9 h 59 min and 10 h 04 min, respectively. Another team that successfully traversed was TerraMax, which scored 27 h and 15 min. For the 2007 competition, which required vehicles to be able to drive fully automatically in urban road environments and obey traffic rules, Carnegie Mellon University's Boss driverless car was the winner.

In 2008, the National Natural Science Foundation of China (NSFC) proposed a major research program on "Cognitive Computing of Audiovisual Information," which started a boom in domestic research on intelligent driving technology. Since 2009, the National Natural Science Foundation of China has hosted the "Future Challenge" every year. The purpose is to use the competition as the carrier to encourage major domestic universities and research institutes to integrate and innovate to develop artificial intelligence and intelligent driving technology. The competition takes the development of an intelligent driving vehicle verification platform with natural environment perception and intelligent behavior decision-making capabilities as the main form, to promote technical seminars and industrial applications, by testing the research results through autonomous driving in a real

Fig. 1.4 The closed test site of the autonomous vehicle in Changshu

road environment. Starting with the fifth competition in 2013, it was held in Changshu, Jiangsu Province, and included two parts: a suburban road test and an urban road test to evaluate the completion of testing tasks by driverless vehicles according to 4S standards (i.e., Safety, Smartness, Smoothness, and Speed). The road environment is more complex and diverse, adding arch bridges, tunnels, ramp entrances, and school entrances to the commonly encountered obstacle cars, slow-moving cars, and temporary road blockage scenarios in the driving process. The ability to intelligently perceive traffic signs, pedestrians, vehicles, and objects and the ability to control autonomous decision-making and correct behavior are the focus of the assessment.

In particular, for the tenth edition of the competition in 2020, the validation of a mixed manned and unmanned driving test was performed for the first time, where multiple unmanned vehicles interact with multiple manned vehicles. As shown in Fig. 1.4, more than ten test scenarios are set up in the nine-grid test area, including traffic markings and signs, traffic signals, non-motorized mixed road sections, pedestrian avoidance sections, construction road closed turnaround sections, construction bypass obstacle sections, simulated tunnel sections, simulated rain sections, rural simple road sections, and real traffic convergence. Compared with highways, urban roads require participating teams to consider interaction and decision-making capabilities in complex scenarios of autonomous driving vehicles. On the basis of improving the open-source digital map of the test area, the semantic topology map was introduced for the first time to complete the natural interactive navigation application test. Meanwhile, with autonomous travel services as the background, the technical maturity of driverless commercial applications has been fully verified.

1.7 Summary

The main key technologies of autonomous vehicles include environment perception, path planning, driving cognition, decision-making, and control. As the first part of autonomous driving, the environment perception system is equipped with onboard sensors to collect road environment information inside and around the vehicle. Onboard sensors have been able to provide rich perception data for autonomous driving, but they still have some limitations that prevent them from achieving full-scene applications. For instance, vision sensors have poor adaptability to changes in light, ultrasonic RaDAR has limited detection range, and the high cost of LiDAR with poor adaptability to disturbances such as rain, snow, and fog.

Due to their respective limitations, single-modal sensing is difficult to completely solve the problem of accurate perception of objects in complex traffic environments. To ensure the integrity of collected information and driving safety, two or more multi-source heterogeneous sensors are often installed on autonomous driving vehicles. The environment perception method based on multimodal data can effectively improve the completeness of autonomous driving data and the accuracy of environment perception, which should be studied in depth. Environmental perception tasks involve target classification, recognition, and tracking, and the difficulties of different task solutions are various. Aiming at the above difficulties, this book focuses on the research of environment perception methods based on the characteristics of multimodal data. Based on the automatic joint calibration algorithm, the study of object classification, recognition, and tracking based on multimodal data features provides a theoretical basis and feasible methods for the perception system of intelligent driving cars, which has great theoretical and practical significance.

References

1. Arnold, E., Al-Jarrah, O.Y., Dianati, M., Fallah, S., Oxtoby, D., Mouzakitis, A.: A survey on 3d object detection methods for autonomous driving applications. IEEE Trans. Intell. Transp. Syst. **20**(10), 3782–3795 (2019)
2. Arras, K.O., Grzonka, S., Luber, M., Burgard, W.: Efficient people tracking in laser range data using a multi-hypothesis leg-tracker with adaptive occlusion probabilities. In: 2008 IEEE International Conference on Robotics and Automation, pp. 1710–1715. IEEE (2008)

3. Bai, M., Mattyus, G., Homayounfar, N., Wang, S., Lakshmikanth, S.K., Urtasun, R.: Deep multi-sensor lane detection. In: 2018 IEEE/RSJ International Conference on Intelligent Robots and Systems (IROS), pp. 3102–3109. IEEE (2018)
4. Bhat, G., Danelljan, M., Gool, L.V., Timofte, R.: Learning discriminative model prediction for tracking. In: Proceedings of the IEEE/CVF International Conference on Computer Vision, pp. 6182–6191 (2019)
5. Bolme, D.S., Beveridge, J.R., Draper, B.A., Lui, Y.M.: Visual object tracking using adaptive correlation filters. In: 2010 IEEE Computer Society Conference on Computer Vision and Pattern Recognition, pp. 2544–2550. IEEE (2010)
6. Eitel, A., Springenberg, J.T., Spinello, L., Riedmiller, M., Burgard, W.: Multimodal deep learning for robust RGB-D object recognition. In: 2015 IEEE/RSJ International Conference on Intelligent Robots and Systems (IROS), pp. 681–687. IEEE (2015)
7. Feng, D., Haase-Schütz, C., Rosenbaum, L., Hertlein, H., Glaeser, C., Timm, F., Wiesbeck, W., Dietmayer, K.: Deep multi-modal object detection and semantic segmentation for autonomous driving: datasets, methods, and challenges. IEEE Trans. Intell. Transp. Syst. **22**(3), 1341–1360 (2020)
8. Hariharan, B., Arbeláez, P., Girshick, R., Malik, J.: Simultaneous detection and segmentation. In: European Conference on Computer Vision, pp. 297–312. Springer (2014)
9. Jafari, O.H., Mitzel, D., Leibe, B.: Real-time RGB-D based people detection and tracking for mobile robots and head-worn cameras. In: 2014 IEEE International Conference on Robotics and Automation (ICRA), pp. 5636–5643. IEEE (2014)
10. Kang, K., Li, H., Yan, J., Zeng, X., Yang, B., Xiao, T., Zhang, C., Wang, Z., Wang, R., Wang, X., et al.: T-cnn: Tubelets with convolutional neural networks for object detection from videos. IEEE Trans. Circuits Syst. Video Technol. **28**(10), 2896–2907 (2017)
11. Linder, T., Breuers, S., Leibe, B., Arras, K.O.: On multi-modal people tracking from mobile platforms in very crowded and dynamic environments. In: 2016 IEEE International Conference on Robotics and Automation (ICRA), pp. 5512–5519. IEEE (2016)
12. Linder, T., Girrbach, F., Arras, K.O.: Towards a robust people tracking framework for service robots in crowded, dynamic environments. In: Assistance and Service Robotics Workshop (ASROB-15) at the IEEE/RSJ Int. Conf. on Intelligent Robots and Systems (IROS) (2015)
13. Luo, J., Yang, Y.: An overview of target detection methods based on data fusion. Control Decision **35**(1), 1–15 (2020)
14. Luo, W., Xing, J., Milan, A., Zhang, X., Liu, W., Kim, T.K.: Multiple object tracking: a literature review. Artif. Intell. **293**, 103,448 (2021)
15. Mees, O., Eitel, A., Burgard, W.: Choosing smartly: adaptive multimodal fusion for object detection in changing environments. In: 2016 IEEE/RSJ International Conference on Intelligent Robots and Systems (IROS), pp. 151–156. IEEE (2016)
16. Xu, Y., Zhou, X., Chen, S., Li, F.: Deep learning for multiple object tracking: a survey. IET Comput. Vis. **13**(4), 355–368 (2019)
17. Yu, J., Jiang, Y., Wang, Z., Cao, Z., Huang, T.: Unitbox: an advanced object detection network. In: Proceedings of the 24th ACM International Conference on Multimedia, pp. 516–520 (2016)
18. Zhang, L., Danelljan, M., Gonzalez-Garcia, A., van de Weijer, J., Shahbaz Khan, F.: Multi-modal fusion for end-to-end rgb-t tracking. In: Proceedings of the IEEE/CVF International Conference on Computer Vision Workshops, pp. 0–0 (2019)
19. Zhang, X., Ye, P., Leung, H., Gong, K., Xiao, G.: Object fusion tracking based on visible and infrared images: a comprehensive review. Inf. Fusion **63**, 166–187 (2020)
20. Zou, Z., Shi, Z., Guo, Y., Ye, J.: Object detection in 20 years: a survey. Preprint (2019). arXiv:1905.05055

Chapter 2
Overview of Data Fusion in Autonomous Driving Perception

Abstract In autonomous driving, research on data fusion has influential academic and application value. This chapter is proposed to summarize the data fusion methods of autonomous driving in recent years. Firstly, the development of deep object detection and data fusion in autonomous driving is introduced, as well as existing reviews. From three aspects of multimodal object detection, fusion levels, and calculation methods, the cutting-edge progress in this field is comprehensively shown. Finally, open issues are discussed, and the performance, challenges, and prospects are summarized.

2.1 A Brief Review of Deep Learning

Convolutional neural networks (CNNs) are one of the most efficient and powerful deep learning models for image processing and understanding. Compared to the multi-layer perceptron (MLP), the CNN is shift-invariant, contains fewer weights, and exploits hierarchical patterns, making it highly efficient for image semantic extraction. The end-to-end trained hidden layers of a CNN consist of convolutional layers, batch normalization layers, activation layers, and pooling layers. This hierarchical structure extracts image features with increasing abstract levels and receptive fields, enabling the learning of high-level semantics.

The point cloud is a set of data points, which are LiDAR's measurements of the detected object's surface. In terms of data structure, the point cloud is sparse, irregular, orderless, and continuous. Point cloud encodes information in 3D structures and in per-point features (reflective intensities, color, normal, etc.), which is invariant to scale, rigid transformation, and permutation. These characteristics made feature extractions on the point cloud challenging for existing deep learning models, which require the modifications of existing models or developing new models. Therefore, this section focuses on introducing common methodologies for point cloud processing.

Volumetric representation based: The volumetric representation partitions the point cloud into a fixed-resolution 3D grid, where features of each grid/voxel are hand-crafted or learned. This representation is compatible with standard 3D

X. Zhang et al., *Multi-sensor Fusion for Autonomous Driving*,
https://doi.org/10.1007/978-981-99-3280-1_2

convolution. Several techniques have been proposed in [21] to reduce overfittings and orientation sensitivity and capture internal structures of objects. However, the volumetric representation loses spatial resolution and fine-grained 3D geometry during voxelization which limits its performance. Furthermore, attempts to increase its spatial resolution (denser voxels) would cause computation and memory footprint to grow cubically, making it unscalable.

Index/tree representation based: To alleviate constraints between high spatial resolution and computational costs, adapted-resolution partition methods that leverage tree-like data structures, such as k-d tree [9] and octree [11], are proposed. By dividing the point cloud into a series of unbalanced trees, regions can be partitioned based on their point densities. This allows regions with lower point densities to have lower resolutions, which reduce unnecessary computation and memory footprint. Point features are extracted along with the pre-built tree structure.

2D view representation based: 2D views/multi-views are generated by projecting the point cloud to multiple 2D view planes. These rendered multi-view images can be processed by standard 2D convolutions, and features from these views are aggregated via view-pooling layers. Thus, the permutation-invariant problem is solved by transforming point cloud to images and the translation-invariant is achieved by aggregating features from different views. Qi et al. [22] combined the volumetric representation with multi-views generated via sphere rendering. Unfortunately, 2D view methods lose the 3D geometry information during the view rendering and struggle with per-point label prediction.

Graph representation based: Point clouds can be represented as graphs, and convolution-like operation can be implemented on graphs in the spatial or spectral domain. For graph convolution in the spatial domain, operations are carried out by MLPs on spatially neighboring points. Spectral-domain graph convolutions extend convolutions as spectral filtering on graphs through the Laplacian spectrum.

Point representation based: Point representation-based methods consume the point cloud without transforming it into an intermediate data representation. Early works in this direction employ shared multi-layer perceptrons (MLPs) to process point cloud [22], while recent works concentrated on defining specialized convolution operations for points [8].

One of the pioneering works of direct learning on point clouds is the PointNet [20], which employs an independent T-Net module to align point clouds and shared MLPs to process individual points for per-point feature extraction. The computation complexity of the PointNet increases linearly with the number of inputs, making it more scalable compared with volumetric-based methods. To achieve permutation invariance, point-wise features are extracted by shared MLPs which are identical for all points. These features are aggregated by symmetric operations (i.e., max pooling), which are also permutation invariant. The feature extraction process of the PointNet is defined as:

$$G(\{x_1, \cdots, x_n\}) \approx f_{sym}(h(x_1), \cdots, h(x_n)) \tag{2.1}$$

where x represents input points, h represents the per-point feature extraction function (i.e., shared MLPs), f_{sym} represents a symmetric function (i.e., max pooling), and g is a general function that we want to approximate.

However, the PointNet fails to extract local interpoint geometry at different levels. To mitigate this challenge, Qi et al. [22] extended the PointNet to extract features from different levels by grouping points into multiple sets and apply PointNets locally. To reduce the computational and memory cost of the PointNet++, the RandLA-Net stacked the random point sampling modules and attention-based local feature aggregation modules hierarchically to progressively increase receptive field while maintaining high efficiency. Unlike PointNet-based methods, the spatial relationship between points is explicitly modeled in point-wise convolutions. Point-wise convolutions aim to generalize the standard 2D discrete convolution to the continuous 3D space. The main challenge is to replace the discrete weight filter in standard convolution with a continuous weight function. This continuous weight function is approximated using MLPs in PointConv [26] and correlation functions in KPConv [24] and PCNN [1]. More specifically, the PCNN defines convolution kernels as 3D points with weights. A Gaussian correlation function that takes the coordinates of the kernel point and input point is used to calculate the weighting matrix at any given 3D coordinates. The KPConv follows this idea but instead uses a linear correlation function. Furthermore, KPConvs are applied on local point patches hierarchically, which are similar to the concepts of standard CNNs. This general point-wise convolution C at an input point $x \in \mathbb{R}^3$ in 3D continuous space is defined as:

$$(F * h)(x) = \sum_{x_i \in N_x} h(x_i - x) f_i \tag{2.2}$$

where h is the per-point kernel function which calculates the weighting matrix given the coordinates of input points and kernel points. x_i and f_i are the ith neighboring points of x and their corresponding features (intensity, color, etc.). N_x are all the neighboring points of the input point x, which are determined using k-NN or radius neighborhoods.

2.2 Fusion in Depth Completion

Depth completion aims to up-sample sparse irregular depth to dense regular depth, which facilitates the downstream perception module. Depth completion can reduce the drastic uneven distributions of points in a LiDAR scan. For instance, far-away objects represented by a hand full of points are up-sampled to match their closer counterparts. To achieve this, high-resolution images are often employed to guide the 3D depth up-sampling. The depth completion task can be represented as:

$$\omega^* = \arg_{\omega} \min \zeta(f(x; \omega), G) \tag{2.3}$$

where the network $f(.)$ parametrized by ω predicts the ground truth G, given the input x. The loss function is represented as $\zeta(., .)$.

The idea behind image-guided depth completion is that dense RGB/color information contains relevant 3D geometry. Therefore, images can be leveraged as a reference for depth up-sampling.

Signal-level fusion: In 2018, Ma et al. [14] presented a ResNet [7]-based autoencoder network that leverages RGB-D images (i.e., images concatenated with sparse depth maps) to predict dense depth maps. However, this method requires pixel-level depth ground truth, which is difficult to obtain. To solve this issue, Ma et al. presented a model-based self-supervised framework that only requires a sequence of images and sparse depth images for training. This self-supervision is achieved by employing sparse depth constrain, photometric loss, and smoothness loss. However, this approach assumes objects to be stationary. Furthermore, the resulting depth output is blurry, and input depth may not be preserved. To generate a sharp dense depth map in real time, Cheng et al. fed RGB-D images to a convolutional spatial propagation network (CSPN). This CSPN aims to extract the image-dependent affinity matrix directly, producing significantly better results in key measurements with lesser runtime. In CSPN++, Cheng et al. proposed to dynamically select convolutional kernel sizes and iterations to reduce computation. Furthermore, CSPN++ employs weighted assembling to boost its performance.

Feature-level fusion: Jaritz et al. presented an autoencoder network that can perform either depth completion or semantic segmentation from sparse depth maps and images without applying validity masks. Images and sparse depth maps are first processed by two parallel NASNet-based encoders [32] before fusing them into the shared decoder. This approach can achieve decent performance with very sparse depth inputs (eight-channel LiDAR). Wang et al. designed an integrable module (PnP) that leverages the sparse depth map to improve the performance of existing image-based depth prediction networks. This PnP module leverages gradient calculated from sparse depth to update the intermediate feature map produced by the existing depth prediction network. El-desokey et al. presented a framework for unguided depth completion that processes images and very sparse depth maps in parallel and combine them in a shared decoder. Furthermore, normalized convolutions are used to process highly sparse depth and to propagate confidence. Valada et al. extended one-stage feature-level fusion to multiple stages of varying depths of the network. Similarly, GuideNet [23] fuse image features to sparse depth features at different stages of the encoder to guide the up-sampling of sparse depths, which achieves top performance in the KITTI depth completion benchmark. The constraint of these approaches is the lack of large-scale datasets that have dense depth ground truth.

Multi-level fusion: Van Gansbeke et al. further combine signal-level fusion and feature-level fusion in an image-guided depth completion network. The network consists of a global and a local branch to process RGB-D data and depth data in parallel before fusing them based on the confidence maps.

Compared with the RGB image, dense depth disparity from stereo cameras contains richer ground truth 3D geometry. On the other hand, LiDAR depth is sparse but of higher accuracy. These complementary characteristics enable stereo-LiDAR fusion-based depth completion models to produce a more accurate dense depth. However, it is worth noting that stereo cameras have limited range and struggles in high-occlusion, texture-less environments, making them less ideal for autonomous driving.

Feature-level fusion: One of the pioneering works is from Park et al. [18], in which high-precision dense disparity map is computed from dense stereo disparity and point cloud using a two-stage CNN. The first stage of the CNN takes LiDAR and stereo disparity to produce a fused disparity. In the second stage, this fused disparity and left RGB image are fused in the feature space to predict the final high-precision disparity. Finally, the 3D scene is reconstructed from this high- precision disparity. The bottleneck of this approach is the lack of large-scale annotated stereo-LiDAR datasets. The LidarStereoNet [4] averted this difficulty with an unsupervised learning scheme, which employs image warping/photometric loss, sparse depth loss, smoothness loss, and plane fitting loss for end-to-end training. Furthermore, the introduction of "feedback loop" makes the LidarStereoNet robust against noisy point cloud and sensor misalignment. Similarly, Zhang et al. presented a self-supervised scheme for depth completion. The loss function consists of sparse depth, photometric, and smoothness loss.

2.3 Fusion in Dynamic Object Detection

Object detection (3D) aims to locate, classify, and estimate oriented bounding boxes in the 3D space. This section is devoted to dynamic object detection, which includes common dynamic road objects (car, pedestrian, cyclist, etc.). There are two main approaches for object detection: sequential and one-step. Sequential-based models consist of a proposal stage and a 3D bounding box (bbox) regression stage in the chronological order. In the proposal stage, regions that may contain objects of interest are proposed. In the bbox regression stage, these proposals are classified based on the region-wise features extracted from 3D geometry. However, the performance of sequential fusion is limited by each stage. On the other hand, one-step models consist of one stage, where 2D and 3D data are processed in a parallel manner.

A 2D proposal-based sequential model attempts to utilize 2D image semantics in the proposal stage, which takes advantage of off-the-shelf image processing models. Specifically, these methods leverage the image object detector to generate 2D region proposals, which are projected to the 3D space as detection seeds. There are two projection approaches to translate 2D proposals to 3D. The first one is projecting bounding boxes in the image plane to the point cloud, which results in a frustum-shaped 3D search space. The second method projects the point cloud to the image plane, which results in the point cloud with point-wise 2D semantics.

Result-level fusion: The intuition behind result-level fusion is to use off-the-shelf 2D object detectors to limit the 3D search space for 3D object detection, which significantly reduces computation and improves runtime. However, since the whole pipeline depends on the results of the 2D object detector, it suffers from the limitations of the image-based detector.

One of the early works of result-level fusion is the F-PointNets [19], where 2D bounding boxes are first generated from images and projected to the 3D space. The resulting projected frustum proposals are fed into a PointNet-based detector for 3D object detection. Du et al. extended the 2D to 3D proposal generation stage with an additional proposal refinement stage, which further reduces unnecessary computation on the background point. During this refinement stage, a model fitting-based method is used to filter out background points inside the seed region. Finally, the filtered points are fed into the bbox regression network. The RoarNet followed a similar idea, but instead employs a neural network for the proposal refinement stage. Multiple 3D cylinder proposals are first generated based on each 2D bbox using the geometric agreement search, which results in smaller but more precise frustum proposals then the F-PointNet. These initial cylinder proposals are then processed by a PointNet-based header network for the final refinement. To summarize, these approaches assume each seed region only contains one object of interest, which is however not true for crowded scenes and small objects like pedestrians.

One possible solution toward the aforementioned issues is to replace the 2D object detector with 2D semantic segmentation and region-wise seed proposal with point-wise seed proposals. Intensive point-based object detector (IPOD) by Yang et al. is a work in this direction. In the first step, 2D semantic segmentation is used to filter out background points. This is achieved by projecting points to the image plane and associated point with 2D semantic labels. The resulting foreground point cloud retains the context information and fine-grained location, which is essential for the region-wise proposal and bbox regression. In the following point-wise proposal generation and bbox regression stage, two PointNet++-based networks are used for proposal feature extraction and bbox prediction. In addition, a novel criterion called PointsIoU is proposed to speed up training and inference. This approach has yielded significant performance advantages over other state-of-the-art approaches in scenes with high occlusion or many objects.

Multi-level fusion: Another possible direction of improvement is to combine result-level fusion with feature-level fusion, where one such work is PointFusion [29]. The PointFusion first utilizes an existing 2D object detector to generate 2D bboxes. These bboxes is used to select corresponding points, via projecting points to the image plane, and locate points that pass through the bboxes. Finally, a ResNet- and a PointNet-based network combine image and point cloud features to estimate 3D objects. In this approach, image features and point cloud features are fused per proposal for final object detection in 3D, which facilitates 3D bbox regression. However, its proposal stage is still amodal. In SIFRNet, frustum proposals are first generated from an image. Point cloud features in these frustum proposals are then combined with their corresponding image features for final 3D bbox regression. To achieve scale invariance, the PointSIFT is incorporated into the network. Additionally, the SENet module is used to suppress less informative features.

Feature-level fusion: Early attempts of multimodal fusion are done in pixel-wise, where 3D geometry is converted to image format or appended as additional channels of an image. The intuition is to project 3D geometry onto the image plane and leverage mature image processing methods for feature extraction. The resulting output is also on the image plane, which is not ideal to locate objects in the 3D space. In 2014, Gupta et al. proposed Deep R-CNN, an R-CNN-based architecture for 2D object detection, instance segmentation, and semantic segmentation. It encodes 3D geometry from Microsoft Kinect camera in image's RGB channels, which are horizontal disparity, height above ground, and angle with gravity (HHA). Gupta et al. extended Depth-RCNN in 2015 for 3D object detection by aligning 3D CAD models, yielding significant performance improvement. In 2016, Gupta et al. developed a novel technique for supervised knowledge transfer between networks trained on image data and unseen paired image modality (depth image). In 2016, Schlosser et al. further exploited learning RGB-HHA representations on 2D CNNs for pedestrian detection. However, the HHA data are generated from the LiDAR's depth instead of a depth camera. The authors also noticed that better results can be achieved if the fusion of RGB and HHA happens at deeper layers of the network.

The resolution mismatch between dense RGB and sparse depth means only a small portion of pixels have corresponding points. Therefore, to directly append RGB information to points leads to the loss of most texture information, rendering the fusion pointless. To mitigate this challenge, PointPainting [25] extract high-level image semantic before the per-point fusion. To be more specific, PointPainting follows the idea of projecting points to 2D semantic maps in [30]. But instead of using 2D semantics to filter non-object points, 2D semantics is simply appended to point clouds as additional channels. The authors argued that this technique made PointPainting flexible as it enables any point cloud networks to be applied on this fused data. To demonstrate this flexibility, the fused point cloud is fed into multiple existing point cloud detectors, which are based on the PointRCNN, the VoxelNet, and the PointPillars. However, this would lead to the coupling between image and LiDAR models. This requires the LiDAR model to be re-trained when the image model changes, which reduces the overall reliability and increases the development cost.

In a 3D proposal-based sequential model, 3D proposals are directly generated from 2D or 3D data. The elimination of 2D to 3D proposal transformation greatly limits the 3D search space for 3D object detection. Common methods for 3D proposal generation include the multi-view approach and the point cloud voxelization approach.

Multi-view-based approach exploits the point cloud's bird's-eye view (BEV) representation for 3D proposal generation. The BEV is the preferred viewpoint because it avoids occlusions and retains the raw information of objects' orientation and x, y coordinates. These orientation and x, y coordinates' information are critical for 3D object detection while making coordinate transformation between BEV and other views straightforward.

This makes it possible to apply standard 3D discrete convolution and leverage existing network structures to process point cloud. The drawback is the loss of some spatial resolution, which might contain fine-grained 3D structure information.

Feature-level fusion: One of the pioneering and most important works in generating 3D proposals from BEV representations is MV3D [3]. MV3D generate 3D proposals on pixelized top-down LiDAR feature map (height, density, and intensity). These 3D candidates are then projected to the LiDAR front view and image plane to extract and fuse region-wise features for bbox regression. The fusion happens at the region of interest (ROI) level via ROI pooling. The ROIviews of views is defined as:

$$ROI_{views} = T_{3D \rightarrow views}(p_{3D}), views \in \{BV, FV, RGB\} \tag{2.4}$$

where $T_{3D \rightarrow views}$ represents the transformation function that project point cloud pan from 3D space to bird's-eye view (BEV), front view (FV), and the image plane (RGB). The ROI pooling R to obtain feature vector fviews is defined as:

$$f_{views} = R(x, ROI_{views}), views \in \{BV, FV, RGB\} \tag{2.5}$$

There are a few drawbacks of the MV3D. Firstly, generating 3D proposals on BEV assumes that all objects of interest are captured without occlusions from this viewpoint. This assumption does not hold well for small object instances, such as pedestrians and bicyclists, which can be fully occluded by other large objects in the point cloud. Secondly, spatial information of small object instances is lost during the down-sample of feature maps caused by consecutive convolution operations. Thirdly, object-centric fusion combines feature maps of image and point clouds through ROI pooling, which spoils fine-grained geometric information in the process. It is also worth noting that redundant proposals lead to repetitive computation in the bbox regression stage. To mitigate these challenges, multiple methods have been put forward to improve MV3D.

To improve the detection of small objects, the Aggregate View Object Detection (AVOD) network [10] first improved the proposal stage in MV3D with feature maps from both BEV point cloud and image. Furthermore, an autoencoder architecture is employed to up-sample the final feature maps to its original size. This alleviates the problem that small objects might get down-sampled to one "pixel" with consecutive convolution operations. The proposed feature fusion Region Proposal Network (RPN) first extracts equal-length feature vectors from multiple modalities (BEV point cloud and image) with crop and resize operations, followed by a 1×1 convolution operation for feature space dimensionality reduction, which can reduce computational cost and boost up speed. Lu et al. also utilized an encoder-decoder-based proposal network with Spatial-Channel Attention (SCA) module and Extension Spatial Upsample (ESU) module. The SCA can capture multi-scale contextual information, whereas ESU recovers the spatial information. One of the problems in object-centric fusion methods is the loss of fine-grained geometric information during ROI pooling. The ContFuse by Liang et al. tackles this information loss with point-wise fusion. This point-wise fusion is achieved with continuous convolutions fusion layers that bridge image and point cloud features of different scales at multiple stages in the network. This is achieved by first extracting k-nearest neighbor points for each pixel in the BEV representation of point cloud. These points are then projected to the image plane to retrieve related image features. Finally, the fused feature vector is weighted according to their geometry offset to the target "pixel" before feeding into MLPs. However, point-wise fusion might fail to take full advantage of high-resolution images when the LiDAR points are sparse. In [12], Liang et al. further extended point-wise fusion by combining multiple fusion methodologies, such as signal-level fusion (RGB-D), feature-level fusion, multi-view, and depth completion. In particular, depth completion up-samples sparse depth map using image information to generate a dense pseudo-point cloud. This up-sampling process alleviates the sparse point-wise fusion problem, which facilitates the learning of cross-modality representations. Furthermore, the authors argued that multiple complementary tasks (ground estimation, depth completion, and 2D/3D object detection) could assist the network achieve better overall performance. However, point-wise/pixel-wise fusion leads to the "feature blurring" problem. This "feature blurring" happens when one point in the point cloud is associated with multiple pixels in the image or the other way around, which confound the data fusion. Similarly, Wang et al. replace the ROI pooling in MV3D with sparse non-homogeneous pooling, which enables effective fusion between feature maps from multiple modalities.

MVX-Net presented by Sindagi et al. introduced two methods that fuse image and point cloud data point-wise or voxel-wise. Both methods employ a pre-trained 2D CNN for image feature extraction and a VoxelNet-based network to estimate objects from the fused point cloud. In the point-wise fusion method, the point cloud is first projected to image feature space to extract image features before voxelization and processed by VoxelNet. The voxel-wise fusion method first voxelized the point cloud before projecting non-empty voxels to the image feature space for voxel-/region-wise feature extraction. These voxel-wise features are only appended to their corresponding voxels at a later stage of the VoxelNet. MVX-Net achieved state-of-the-art results and outperformed other LiDAR-based methods on the KITTI benchmark while lowering false-positive and false-negative rate compared to [31].

The simplest means to combine the voxelized point cloud and image is to append RGB information as additional channels of a voxel. In a 2014 paper by Song et al., 3D object detection is achieved by sliding a 3D detection window on the voxelized point cloud. The classification is performed by an ensemble of Exemplar-SVMs. In this work, color information is appended to voxels by projection. Song et al. further extended this idea with 3D discrete convolutional neural networks. In the first stage, the voxelized point cloud (generated from RGB-D data) is first processed by multi-scale 3D RPN for 3D proposal generation. These candidates are then classified by joint Object Recognition Network (ORN), which takes both image and voxelized point cloud as inputs. However, the volumetric representation introduces boundary artifacts and spoils fine-grained local geometry. Secondly, the resolution mismatch between image and voxelized point cloud makes fusion inefficient.

One-step models perform proposal generation and bbox regression in a single stage. By fusing the proposal and bbox regression stage into one step, these models are often more computationally efficient. This makes them more well suited for real-time applications on mobile computational platforms. Meyer et al. extended the LaserNet to multi-task and multimodal network, performing 3D object detection and 3D semantic segmentation on fused image and LiDAR data. Two CNN process depth image (generated from point cloud) and front-view image in a parallel manner and fuse them via projecting points to the image plane to associate corresponding image features. This feature map is fed into the LaserNet to predict per-point distributions of the bounding box and combine them for final 3D proposals. This method is highly efficient while achieving state-of-the-art performance.

2.4 Fusion in Stationary Road Object Detection

This section focuses on reviewing recent advances in camera-LiDAR fusion-based stationary road object detection methods. Stationary road objects can be categorized into on-road objects (e.g., road surfaces and road markings) and off-road objects (e.g., traffic signs). On-road and off-road objects provide regulations, warning bans, and guidance for autonomous vehicles.

Existing surveys [15, 17] have presented detailed reviews on traditional multi-modal road detection methods. These methods [27, 28] mainly rely on vision for road/lane detection while utilizing LiDAR for the curb fitting and obstacle masking. Therefore, this section focuses on recent advances in deep learning-based fusion strategies for road extraction.

Deep leaning-based road detection methods can be grouped into BEV-based or front camera view-based. BEV-based methods project LiDAR depth and images to BEV for road detection, which retains original x, y coordinates and orientation of each object. In [2, 13], the dense BEV height estimation is predicted from the point cloud using a CNN, which is then fused with the BEV image for accurate lane detection. However, this method cannot distinguish different lane types. Similarly, Lv et al. also utilized the BEV LiDAR grid map and the BEV image but instead processed them in a parallel manner. Yu et al. proposed a multi-stage fusion strategy (MSRF) that combines image depth features at different network levels, which significantly improves its performance. However, this strategy also relatively increases its computational cost. Wulff et al. used signal-level fusion to generate a fused BEV occupation grid, which is processed by a U-Net-based road segmentation network. However, the signal-level fusion between dense RGB and sparse depth leads to the loss of dense texture information due to the low grid resolution.

Front camera view-based methods project LiDAR depth to the image plane to extract road surface, which suffers from accuracy loss in the translation of 2D to 3D boundaries. The LCNet compared signal-level fusion (early fusion) and feature-level fusion (late fusion and cross-fusion) for road detection, which finds the cross-fusion is the best performing fusion strategy. Similar to [13], PLARD fuses image and point cloud features progressively in multiple stages. Lee et al. focused on improving speed via a spherical coordinate transformation scheme that reduces the input size. These transformed camera and LiDAR data are further processed by a SegNet-based semantic segmentation network.

In LiDAR scans, traffic signs are highly distinguishable due to its retro-reflective property, but the lack of dense texture makes it difficult to classify. On the contrary, traffic sign image patches can be easily classified. However, it is difficult for vision-based TSR system to locate these traffic signs in the 3D space. Therefore, various studies have proposed to utilize both camera and LiDAR for TSR. Existing reviews [15, 16] have comprehensively covered traditional traffic sign recognition methods and part of the deep learning methods. Hence, this section presents a brief overview of traditional traffic sign recognition methods and mostly focuses on recent advances. In a typical TSR fusion pipeline, traffic signs are first located in the LiDAR scan based on its retro-reflective property. These 3D positions of detected traffic signs are then projected to the image plane to generate traffic sign patches, which are fed into an image classifier for classification.

For methods that employ the typical TSR fusion pipeline, the main difference is on the classifier. These classifiers include deep Boltzmann machine (DBM)-based hierarchical classifier, SVMs, and DNN. To summarize, these methods all employ result-level fusion and a hierarchical object detection model. They assume traffic signs are visible in the LiDAR scan, which sometimes is not the case due to the occlusion. Furthermore, this pipeline is limited by the detection range of mobile LiDARs.

To mitigate these challenges, Deng et al. combined image and point cloud to generate a colorized point cloud for both traffic sign detection and classification. In addition, the 3D geometrical properties of detected traffic signs are utilized to reduce false positives. In [6], traffic signs are detected based on prior knowledge, which includes road geometry information and traffic sign geometry information. The detected traffic sign patches are classified by a Gaussian-Bernoulli DBM model. Following this ideal, Guan et al. further improved the traffic sign recognition part using a convolutional capsule network. To summarize, these methods improve the traffic sign detection stage with multimodal data and prior knowledge. However, prior knowledge is often region-specific, which makes it difficult to generalize to other parts of the world.

2.5 Fusion in Semantic Segmentation

This section reviews existing camera-LiDAR fusion methods for 2D semantic segmentation, 3D semantic segmentation, and instance segmentation. 2D/3D semantic segmentation aims to predict per-pixel and per-point class labels, while instance segmentation also cares about individual instances.

There are several 2D semantic segmentation methods, such as feature-level fusion. Valada et al. employed a multi-stage feature-level fusion of varying depths to facilitate semantic segmentation. Caltagirone et al. utilized up-sampled depth image and image for 2D semantic segmentation. This dense depth image is up-sampled using sparse depth images (from point cloud) and images. The best performing cross-fusion model processes dense depth image and image data in two parallel CNN branches with skip connections in between and fuses the two feature maps in the final convolution layer.

There are also several 3D semantic segmentation methods. Feature-level fusion: Dai et al. presented 3DMV, a multi-view network for 3D semantic segmentation which fuses image semantic and point features in voxelized point cloud. Image features are extracted by 2D CNNs from multiple aligned images and projected back to the 3D space. These multi-view image features are max-pooled voxel-wise and fused with 3D geometry before feeding into the 3D CNNs for per-voxel semantic prediction. 3DMV outperformed other voxel-based approaches on ScanNet benchmark. However, the performance of voxel-based approaches is determined by the voxel resolution and hindered by voxel boundary artifacts.

To alleviate problems caused by point cloud voxelization, Chiang et al. proposed a point-based semantic segmentation framework (UPF) that also enables efficient representation learning of image features, geometrical structures, and global context priors. Features of rendered multi-view images are extracted using a semantic segmentation network and projected to 3D space for point-wise feature fusion. This fused point cloud is processed by two PointNet++-based encoders to extract local and global features before feeding into a decoder for per-point semantic label prediction. Similarly, Multi-View PointNet (MVPNet) fused multi-view images semantics and 3D geometry to predict per-point semantic labels.

Permutohedral lattice representation is an alternative approach for multimodal data fusion and processing. Sparse Lattice Networks (SPLATNet) by Su et al. employed sparse bilateral convolution to achieve spatial-aware representation learning and multimodal (image and point cloud) reasoning. In this approach, point cloud features are interpolated onto a bi-dimensional permutohedral lattice, where bilateral convolution is applied. The results are interpolated back onto the point cloud. Image features are extracted from multi-view images using a CNN and projected to the 3D lattice space to be combined with 3D features. This fused feature map is further processed by CNN to predict the per-point label.

In essence, instance segmentation aims to perform semantic segmentation and object detection jointly. It extends the semantic segmentation task by discriminating against individual instances within a class, which makes it more challenging.

Proposal based: Hou et al. presented 3D-SIS, a two-stage 3D CNN that performs voxel-wise 3D instance segmentation on multi-view images and RGB-D scan data. In the 3D detection stage, multi-view image features are extracted and down-sampled using ENet-based network. This down-sample process tackles the mismatch problem between a high-resolution image feature map and a low-resolution voxelized point cloud feature map. These down-sampled image feature maps are projected back to 3D voxel space and append to the corresponding 3D geometry features, which are then fed into a 3D CNN to predict object classes and 3D bbox poses. In the 3D mask stage, a 3D CNN takes images, point cloud features, and 3D object detection results to predict per-voxel instance labels. Narita et al. extended 2D panoptic segmentation to perform scene reconstruction, 3D semantic segmentation, and 3D instance segmentation jointly on RGB images and depth images. This approach takes RGB and depth frames as inputs for instance and 2D semantic segmentation networks. To track labels between frames, these frame-wise predicted panoptic annotations and corresponding depth are referenced by associating and integrating to the volumetric map. In the final step, a fully connected conditional random field (CRF) is employed to fine-tune the outputs. However, this approach does not support dynamic scenes and are vulnerable to long-term post-drift.

Proposal-free based: Elich et al. presented 3D-BEVIS, a framework that performs 3D semantic and instance segmentation tasks jointly using the clustering method on points aggregated with 2D semantics. 3D-BEVIS first extracts global semantic score map and instance feature map from 2D BEV representation (RGB and height above ground). These two semantic maps are propagated to points using a graph neural network. Finally, the mean shift algorithm uses these semantic features to cluster points into instances. This approach is mainly constraint by its dependence on semantic features from BEV, which could introduce occlusions from sensor displacements.

2.6 Fusion in Object Tracking

Multiple object tracking (MOT) aims to maintain objects' identities and track their location across data frames (over time), which is indispensable for the decision-making of autonomous vehicles. To this end, this section reviews camera-LiDAR fusion-based object tracking methods. Based on object initialization methods, MOT algorithms can be catalogued into detection-based tracking (DBT) and detection-free tracking (DFT) frameworks. DBT or tracking-by-detection framework leverages a sequence of object hypotheses and higher-level cues produced by an object detector to track objects. In DBT, objects are tracked via data (detection sequence) association or multiple hypothesis tracking. On the contrary, the DFT framework is based on finite set statistics (FISST) for state estimation. Common methods include multi-target multi-Bernoulli (MeMBer) filter and probability hypothesis density (PHD) filter.

Detection-based tracking (DBT): The tracking-by-detection framework consists of two stages. In the first stage, objects of interest are detected. The second stage associates these objects over time and formulates them into trajectories, which are formulated as linear programs. Frossard et al. presented an end-to-end trainable tracking-by-detection framework comprised of multiple independent networks that leverage both image and point cloud. This framework performs object detection, proposal matching and scoring, and linear optimization consecutively. To achieve end-to-end learning, detection and matching are formulated via a deep structured model (DSM). Zhang et al. presented a sensor-agnostic framework, which employs a loss-coupling scheme for image and point cloud fusion. Similar to [5], the framework consists of three stages, object detection, adjacency estimation, and linear optimization. In the object detection stage, image and point cloud features are extracted via a VGG-16 and a PointNet in parallel and fused via a robust fusion module. The robust fusion module is designed to work with both amodal and multimodal inputs. The adjacency estimation stage extends min-cost flow to multimodality via adjacent matrix learning. Finally, an optimal path is computed from the min-cost flow graph. Tracking and 3D reconstruct tasks can be performed jointly. Extending this idea, Luiten et al. leveraged 3D reconstruction to improve tracking, making tracking robust against complete occlusion. The proposed MOTS-Fusion consists of two stages. In the first stage, detected objects are associated with spatiotemporal tracklets. These tracklets are matched and merged into trajectories using the Hungarian algorithm. Furthermore, MOTSFusion can work with LiDAR mono and stereo depth.

Detection-free tracking (DFT): In DFT, objects are manually initialized and tracked via filtering-based methods. The Complexer-YOLO is a real-time framework for decoupled 3D object detection and tracking on image and point cloud data. In the 3D object detection phase, 2D semantics are extracted and fused point-wise to the point cloud. This semantic point cloud is voxelized and fed into a 3D Complex-YOLO for 3D object detection. To speed up the training process, IoU is replaced by a novel metric called Scale-Rotation-Translation (SRT) score, which evaluates 3 DoFs of the bounding box position. Multi-object tracking is decoupled from the detection, and the inference is achieved via labeled multi-Bernoulli random finite set (LMB RFS) filter.

2.7 Summary

Data fusion is an important trend in the perception task of autopilot. Therefore, this chapter provides an overview of multimodal data fusion methods in target detection in autopilot scene. This chapter introduces the background of target detection and data fusion and summarizes the current research from two aspects: fusion level and fusion calculation method. We believe that the traditional "front, middle, and back" division method does not distinguish the fusion methods in the integration model. Therefore, this chapter adopts a new hierarchical definition method and gives a clear definition, which can help researchers better understand the motivation and function of fusion method design. In addition, this chapter puts forward the rationality analysis of data fusion, summarizes the existing research, discusses the challenges and strategies of current fusion methods, and puts forward some open problems.

Although new fusion methods have been proposed, there is still a lack of theoretical analysis and in-depth comparative experiments. The development of automatic driving technology depends on efficient and robust environment perception. The corresponding data fusion methods should be developed as soon as possible to ensure the performance of vehicle perception technology. Our follow-up work includes the research on data representation and the information gain measurement of cross-modal data fusion. At the same time, we expect that the new research work can solve the challenges in the fusion methods discussed in this chapter.

References

1. Atzmon, M., Maron, H., Lipman, Y.: Point convolutional neural networks by extension operators. Preprint (2018). arXiv:1803.10091
2. Bai, M., Mattyus, G., Homayounfar, N., Wang, S., Lakshmikanth, S.K., Urtasun, R.: Deep multi-sensor lane detection. In: 2018 IEEE/RSJ International Conference on Intelligent Robots and Systems (IROS), pp. 3102–3109. IEEE (2018)
3. Chen, X., Ma, H., Wan, J., Li, B., Xia, T.: Multi-view 3d object detection network for autonomous driving. In: Proceedings of the IEEE Conference on Computer Vision and Pattern Recognition, pp. 1907–1915 (2017)
4. Cheng, X., Zhong, Y., Dai, Y., Ji, P., Li, H.: Noise-aware unsupervised deep lidar-stereo fusion. In: Proceedings of the IEEE/CVF Conference on Computer Vision and Pattern Recognition, pp. 6339–6348 (2019)
5. Frossard, D., Urtasun, R.: End-to-end learning of multi-sensor 3d tracking by detection. In: 2018 IEEE International Conference on Robotics and Automation (ICRA), pp. 635–642. IEEE (2018)
6. Guan, H., Yan, W., Yu, Y., Zhong, L., Li, D.: Robust traffic-sign detection and classification using mobile lidar data with digital images. IEEE J. Sel. Top. Appl. Earth Obs. Remote Sens. **11**(5), 1715–1724 (2018)
7. He, K., Zhang, X., Ren, S., Sun, J.: Deep residual learning for image recognition. In: Proceedings of the IEEE Conference on Computer Vision and Pattern Recognition, pp. 770–778 (2016)
8. Hua, B.S., Tran, M.K., Yeung, S.K.: Pointwise convolutional neural networks. In: Proceedings of the IEEE Conference on Computer Vision and Pattern Recognition, pp. 984–993 (2018)
9. Klokov, R., Lempitsky, V.: Escape from cells: deep kd-networks for the recognition of 3d point cloud models. In: Proceedings of the IEEE International Conference on Computer Vision, pp. 863–872 (2017)
10. Ku, J., Mozifian, M., Lee, J., Harakeh, A., Waslander, S.L.: Joint 3d proposal generation and object detection from view aggregation. In: 2018 IEEE/RSJ International Conference on Intelligent Robots and Systems (IROS), pp. 1–8. IEEE (2018)
11. Lei, H., Akhtar, N., Mian, A.: Octree guided cnn with spherical kernels for 3d point clouds. In: Proceedings of the IEEE/CVF Conference on Computer Vision and Pattern Recognition, pp. 9631–9640 (2019)
12. Liang, M., Yang, B., Chen, Y., Hu, R., Urtasun, R.: Multi-task multi-sensor fusion for 3d object detection. In: Proceedings of the IEEE/CVF Conference on Computer Vision and Pattern Recognition, pp. 7345–7353 (2019)
13. Lv, X., Liu, Z., Xin, J., Zheng, N.: A novel approach for detecting road based on two-stream fusion fully convolutional network. In: 2018 IEEE Intelligent Vehicles Symposium (IV), pp. 1464–1469. IEEE (2018)
14. Ma, F., Karaman, S.: Sparse-to-dense: depth prediction from sparse depth samples and a single image. In: 2018 IEEE International Conference on Robotics and Automation (ICRA), pp. 4796–4803. IEEE (2018)

15. Ma, L., Li, Y., Li, J., Wang, C., Wang, R., Chapman, M.A.: Mobile laser scanned point-clouds for road object detection and extraction: a review. Remote Sens. **10**(10), 1531 (2018)
16. Mogelmose, A., Trivedi, M.M., Moeslund, T.B.: Vision-based traffic sign detection and analysis for intelligent driver assistance systems: perspectives and survey. IEEE Trans. Intell. Transp. Syst. **13**(4), 1484–1497 (2012)
17. Narote, S.P., Bhujbal, P.N., Narote, A.S., Dhane, D.M.: A review of recent advances in lane detection and departure warning system. Pattern Recogn. **73**, 216–234 (2018)
18. Park, K., Kim, S., Sohn, K.: High-precision depth estimation using uncalibrated lidar and stereo fusion. IEEE Trans. Intell. Transp. Syst. **21**(1), 321–335 (2019)
19. Qi, C.R., Liu, W., Wu, C., Su, H., Guibas, L.J.: Frustum pointnets for 3d object detection from RGB-D data. In: Proceedings of the IEEE Conference on Computer Vision and Pattern Recognition, pp. 918–927 (2018)
20. Qi, C.R., Su, H., Mo, K., Guibas, L.J.: Pointnet: deep learning on point sets for 3d classification and segmentation. In: Proceedings of the IEEE Conference on Computer Vision and Pattern Recognition, pp. 652–660 (2017)
21. Qi, C.R., Su, H., Nießner, M., Dai, A., Yan, M., Guibas, L.J.: Volumetric and multi-view CNNs for object classification on 3d data. In: Proceedings of the IEEE Conference on Computer Vision and Pattern Recognition, pp. 5648–5656 (2016)
22. Qi, C.R., Yi, L., Su, H., Guibas, L.J.: Pointnet++: deep hierarchical feature learning on point sets in a metric space. Preprint (2017). arXiv:1706.02413
23. Tang, J., Tian, F.P., Feng, W., Li, J., Tan, P.: Learning guided convolutional network for depth completion. IEEE Trans. Image Process. **30**, 1116–1129 (2020)
24. Thomas, H., Qi, C.R., Deschaud, J.E., Marcotegui, B., Goulette, F., Guibas, L.J.: Kpconv: flexible and deformable convolution for point clouds. In: Proceedings of the IEEE/CVF International Conference on Computer Vision, pp. 6411–6420 (2019)
25. Vora, S., Lang, A.H., Helou, B., Beijbom, O.: Pointpainting: sequential fusion for 3d object detection. In: Proceedings of the IEEE/CVF Conference on Computer Vision and Pattern Recognition, pp. 4604–4612 (2020)
26. Wu, W., Qi, Z., Fuxin, L.: Pointconv: deep convolutional networks on 3d point clouds. In: Proceedings of the IEEE/CVF Conference on Computer Vision and Pattern Recognition, pp. 9621–9630 (2019)
27. Xiao, L., Dai, B., Liu, D., Hu, T., Wu, T.: Crf based road detection with multi-sensor fusion. In: 2015 IEEE Intelligent Vehicles Symposium (IV), pp. 192–198. IEEE (2015)
28. Xiao, L., Wang, R., Dai, B., Fang, Y., Liu, D., Wu, T.: Hybrid conditional random field based camera-lidar fusion for road detection. Inf. Sci. **432**, 543–558 (2018)
29. Xu, D., Anguelov, D., Jain, A.: Pointfusion: deep sensor fusion for 3d bounding box estimation. In: Proceedings of the IEEE Conference on Computer Vision and Pattern Recognition, pp. 244–253 (2018)
30. Yang, Z., Sun, Y., Liu, S., Shen, X., Jia, J.: Ipod: intensive point-based object detector for point cloud. Preprint (2018). arXiv:1812.05276
31. Zhou, Y., Tuzel, O.: Voxelnet: end-to-end learning for point cloud based 3d object detection. In: Proceedings of the IEEE Conference on Computer Vision and Pattern Recognition, pp. 4490–4499 (2018)
32. Zoph, B., Vasudevan, V., Shlens, J., Le, Q.V.: Learning transferable architectures for scalable image recognition. In: Proceedings of the IEEE Conference on Computer Vision and Pattern Recognition, pp. 8697–8710 (2018)

Part II
Method

This section focuses on different tasks of autonomous driving perception based on multimodal fusion. It consists of four chapters corresponding to sensor calibration, object detection, semantic segmentation, and localization tasks. Each chapter introduces novel technical approaches. In Chap. 3, joint calibration is introduced as a basic technique for multimodal fusion. Taking 3D object detection and semantic segmentation as specific tasks, respectively, Chaps. 4 and 5 propose various fusion methods. Chapter 6, as an extension of perception technology, introduces the localization method based on multi-sensor fusion. These methods improve perception accuracy by fusing information from different modalities, providing a foundation for autonomous vehicle applications.

Chapter 3
Multi-Sensor Calibration

Abstract Reliable real-time extrinsic parameters of 3D Light Detection and Ranging (LiDAR) and cameras are vital components of multimodal perception systems. However, extrinsic transformation may drift gradually during operation, resulting in decreased accuracy of the perception system. To solve this problem, this chapter proposes a line-based method that enables automatic online extrinsic calibration of LiDAR and cameras in real-world scenes. Moreover, adaptive optimization is utilized to provide accurate extrinsic parameters. Experiments verify that the proposed approach automatically corrects miscalibration errors and achieves an accuracy of 0.2° . This work can provide a basis for perception systems and further improve the performance of other algorithms that utilize these sensors.

3.1 Introduction

Multiple LiDARs and cameras are widely applied on autonomous vehicles and robots in many scenarios, such as autonomous driving [9, 13], object classification [7, 11], segmentation [12], and SLAM [5, 15]. LiDARs can provide accurate three-dimensional geometric information but with sparse points. On the contrary, cameras are capable of offering a rich representation of the environment but with less accurate distance information. The extrinsic transformation matrix of these sensors can combine these two complementary sensors and maximize the effects. Therefore, accurate extrinsic parameters between them are crucial to provide accurate information for perception systems.

Traditional manual calibration methods require specially designed objects, such as checkboards [2] or manually selected points [6], which lead to a cumbersome calibration procedure. Furthermore, long-time operation and different loads can result in slight drifting and deviations to extrinsic parameters. Therefore, an automatic online correction is needed to adjust to this unforeseen sensor movement.

Current automatic calibration works utilize mutual information [10] or artificially designed targets [3] to calibrate the extrinsic parameters. The use of specific targets and intensity information limits the calibration process to laboratory settings and specific sensors. Some other feature-based calibration method [1] utilizes edge

X. Zhang et al., *Multi-sensor Fusion for Autonomous Driving*,
https://doi.org/10.1007/978-981-99-3280-1_3

features to compute the extrinsic parameters. However, these features are not well corresponded to each other in some scenarios.

3.2 Line-Based Multi-Sensor Calibration

In this section, the line feature is selected to constrain the extrinsic parameters for its ubiquity. Initially, the line features are extracted and filtered from point clouds and images. Afterward, an adaptive optimization is utilized to provide accurate extrinsic parameters. The proposed methods prove that line features are robust geometric features that can be extracted from point clouds and images, thus contributing to extrinsic calibration. To demonstrate the benefits of this method, we evaluated it on the KITTI benchmark with ground truth value. The experiments verify the accuracy of the calibration approach. In online experiments on hundreds of frames, the proposed approach automatically corrects miscalibration errors and achieves an accuracy of $0.2°$, which verifies its applicability in various scenarios. This work can provide a basis for perception systems and further improve the performance of other algorithms that utilize these sensors.

This work aims to release this limitation and simplify the process required to calibrate the extrinsic parameters of LiDAR and cameras. For this purpose, we utilize the robust and widespread line features to compute the transformation matrix between these sensors automatically with no hand-labeling. As input, the proposed method requires a single image and previous several point clouds as well as the initial extrinsic parameters. The line features used in this work are generally spread in outdoor settings, such as trees, street lamps, cars, etc. By applying these line features, the proposed method can automatically adjust according to drifting and changes of the extrinsic transformation matrix, as shown in Fig. 3.1. The main contributions of this section are as follows:

Firstly, we introduce a novel extrinsic calibration method that can automatically estimate the six degrees of freedom (6-DoF) parameters online. The proposed method utilizes generic line features to reduce the drift errors between sensors without requiring manually selected points and special targets; therefore, it is able to be applied to any given scenario.

Secondly, we present a point cloud line extraction approach that can obtain line features from point clouds, where a point cloud processing approach is utilized to filter out noise data and extract line features accurately.

Thirdly, we introduce an adaptive optimization method and a result confidence estimation method to enable optimization toward the correct direction and computing the calibration results efficiently.

The rest of this chapter is organized as follows: Sect. 3.2 introduces the methods proposed in this chapter and evaluates the accuracy of the proposed methods based on the KITTI dataset. Section 3.3 summarizes our research and prospects for future work.

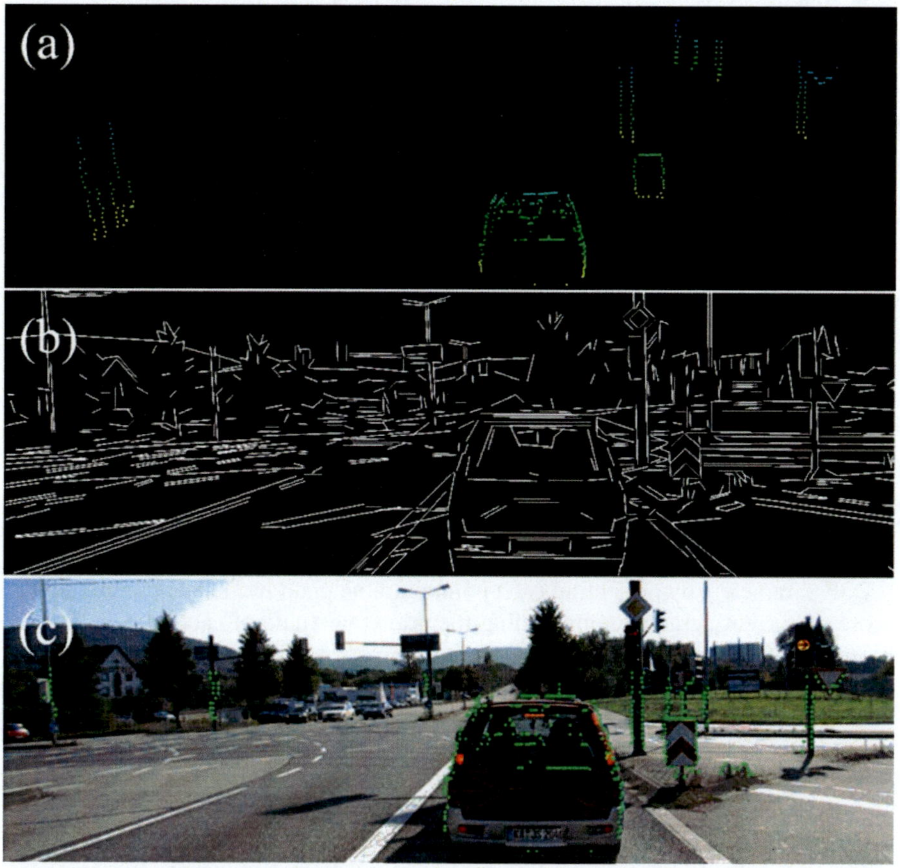

Fig. 3.1 Line features are extracted from a point cloud (**a**) and an image (**b**) of the same scene. By computing the correspondence between these 2D and 3D line features, the projected point cloud line features (**c**) are well aligned with the image line features

3.2.1 Methodology

In this section, line features are selected to accurately calibrate the extrinsic parameters of a LiDAR-camera system. The main benefit of these features is that they are ubiquitous in both outdoor and indoor scenarios. Besides that, they have good correspondence in point clouds and images. Figure 3.2 offers an overview of the proposed approach. The proposed method utilizes three steps. First, a series of preprocessing methods are employed to images and point clouds for further feature extraction. Then, the line features are extracted from images and point clouds and refined by applying feature filtration. Finally, the point cloud line features are projected onto pixel frames by adding small perturbation to initial extrinsic

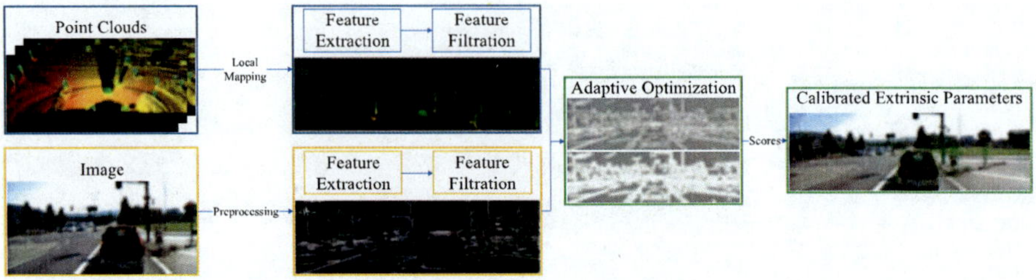

Fig. 3.2 The architecture of the proposed calibration system

parameters, during which the score of each perturbation will be computed and optimized. Details are presented in the following sections.

The extrinsic calibration problem of the LiDAR and camera lies in determining the correct transform matrix between them. In this section, we define the problem as finding the rotation angle vector $\theta = (\theta_x, \theta_y, \theta_z)$ and translation vector $t = (t_x, t_y, t_z)$. For the point clouds and images, we notate the point cloud as P_t and image as I^t_{ij}, representing point i and pixel ij value at frame t. We optimize the 6-DoF parameters by projecting each point onto the pixel frame and computing the score of current parameters by adding the grayscale value of each pixel. The cost score can be computed by projecting LiDAR points onto images, and the objective function can be defined as:

$$S_t = \sum_{t=n-w}^{n} I^t_{ij} \left[\alpha \sum_{p_t}^{p_t \in F^t_h} T_t p_t + (1-\alpha) \sum_{p_t}^{p_t \in F^t_v} T_t p_t \right] / w \tag{3.1}$$

where each LiDAR point p_t iterates over horizontal features F_h and vertical features F_v, respectively. The coefficient α assigns different weights to horizontal and vertical line features. In this section, α is assigned as 0.65 to enhance the constraint of horizontal errors. In addition, w is the size of the sliding window. The score of frame t is computed by considering the previous w frames.

The basic assumption of this work is that the line features from point clouds are highly responsive to line features from images when the calibration parameters are correct. Besides that, it is also assumed that the intrinsic parameters of the camera and LiDAR are already calibrated and the LiDAR data and camera images are captured simultaneously.

In image processing, the RGB images are initially converted into grayscale images, after which the line features are extracted by a line detection algorithm [14]. Afterward, a distance transform model is applied to these gray images. An example of an original image, line features, and edge features is shown in Fig. 3.3. The white edges in Fig. 3.3b and the white lines in Fig. 3.3c represent clustered edge features and line features, respectively. As is shown in Fig. 3.3b, the clustered edge features are more disordered after applying the distance transform model. On the contrary, line features in Fig. 3.3c are more well organized, generating smaller

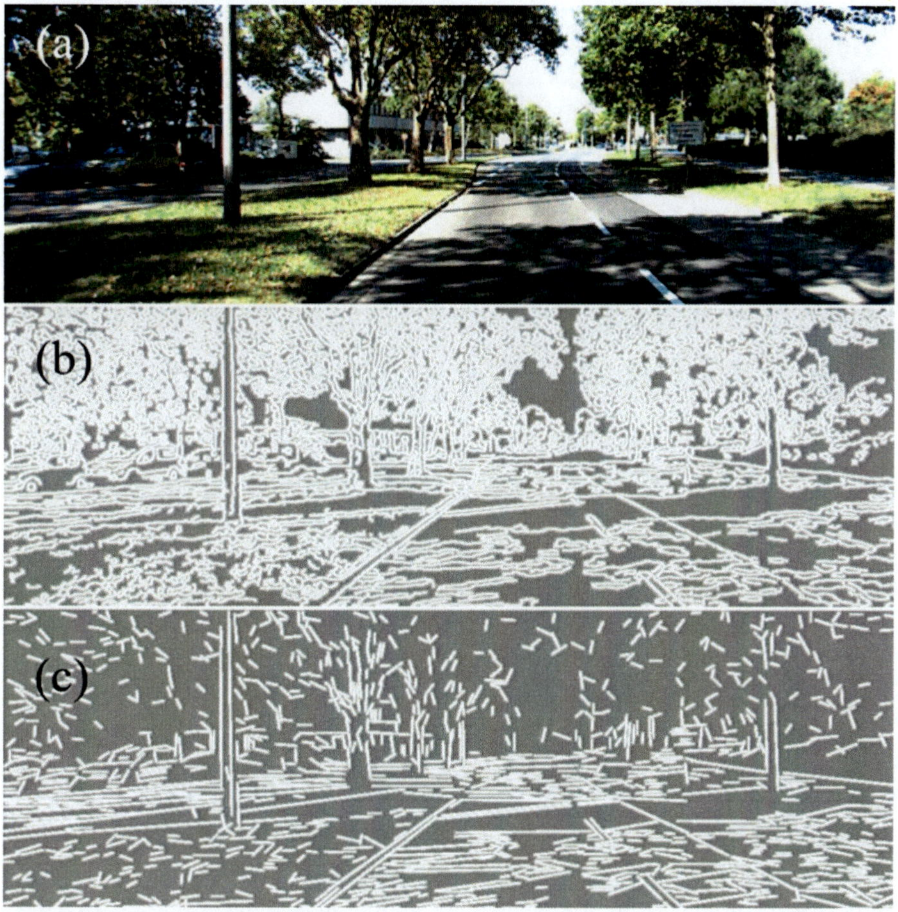

Fig. 3.3 A comparison of edge features and line features. (**a**) shows the original scenario. (**b**) shows the edge features, and (**c**) shows the line features. The white pixels represent the extracted features, and the grayscale changes represent the distance to edge or line features. The whiter the pixels are, the closer they are to the center of these line features

grayscale changes. It can allow for a larger searching step size, thus preventing the optimization process from getting into a local solution.

In LiDAR processing, the principle is to utilize distance discontinuity to get more boundary line features. To achieve this goal, a local mapping method is applied to combine three frames of point cloud into one, which can present more points in one frame. Specifically, a Normal Distribution Transform (NDT) method is utilized to compute the transformation matrix between the current and previous two frames. A comparison between boundary line points extracted in a single frame and a three-in-one frame is shown in Fig. 3.4. Figure 3.4a shows a more intensive point cloud by transforming three frames point cloud $P_{t-2:t}$ to one frame P_t, which can reveal more

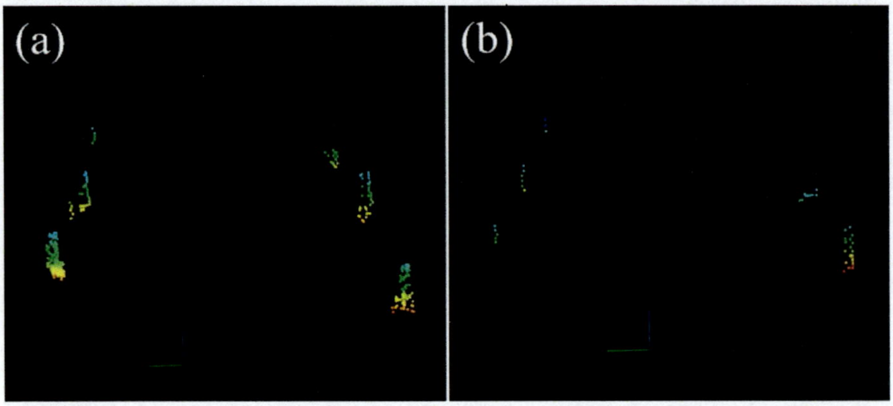

Fig. 3.4 A comparison of line features from a three-in-one point cloud (**a**) and a single point cloud (**b**)

points compared with the other one in Fig. 3.4b. This can improve the extraction performance, especially when a low-beam LiDAR is applied.

Afterward, the denser point cloud is transformed into image form, with each pixel storing the distance information of the corresponding LiDAR point. By comparing the distance between current point and adjacent points, more accurate line features can be extracted by eliminating outliers that are too far away from neighbors. It should be noticed that different from [8] which considered each beam independently, we take the distance information between multiple beams into consideration. It allows the proposed method to extract horizontal features, thus minimizing both horizontal and vertical errors using the line features. The horizontal line features F^h and vertical line F^v features are stored in two different point clouds, respectively. In this setting, the plane intersection lines, a rarely appeared feature, are neglected, which is beneficial to enhance the computation efficiency.

The extracted line features from point clouds can be disordered after the previous section. Therefore, two filtration methods are employed to eliminate outliers. Since the point clouds are already transformed into image form, a convolution kernel is designed to filter out points that are far away from all eight adjacent points beyond a certain threshold. The line features before and after filtration can be seen in Fig. 3.5. This filtration method can remove all the outliers as well as the points that correspond to the ground. As a result, the remaining features can be identified as line features.

After the first filtration, a point cloud cluster algorithm is applied to remove line features that have few adjacent points. The above two filtration steps can provide more well-organized point cloud line features, which guarantee a better optimization result in the subsequent steps.

A more organized line feature from images can be obtained after filtering outline features that are shorter than eight pixels.

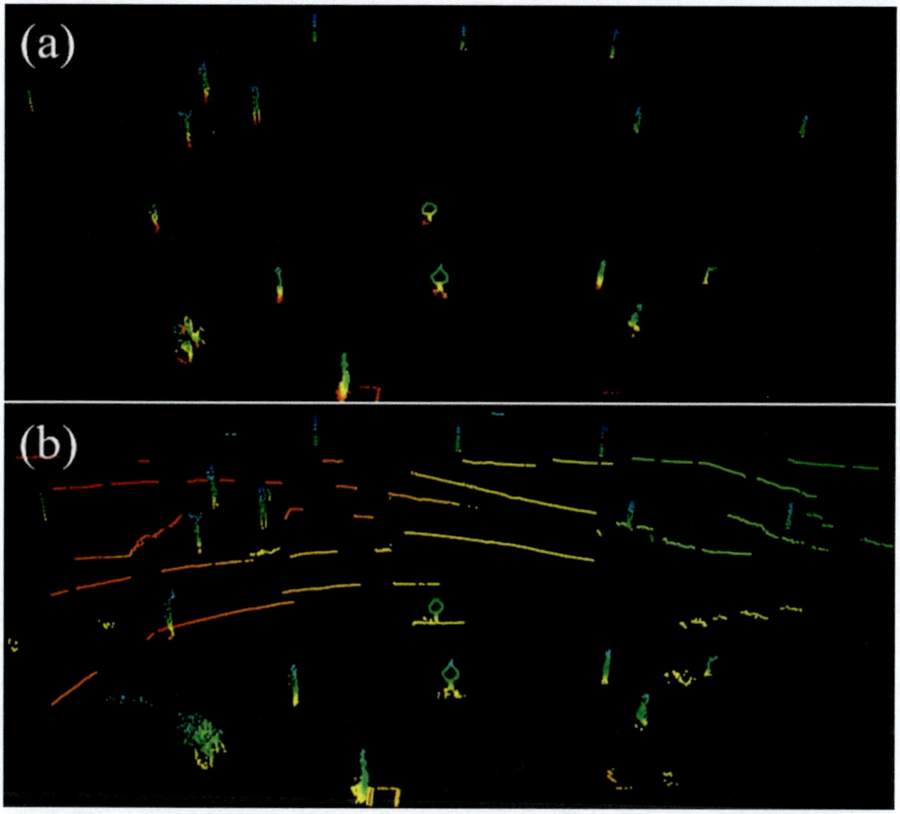

Fig. 3.5 A comparison of line features after (**a**) and before (**b**) feature filtration

In the optimization process, the proposed approach takes both computation accuracy and efficiency into consideration. Before optimization, the proposed approach project extracted LiDAR line features onto the image, and the proportion of the LiDAR points projected onto the gray area will be computed.

For computation accuracy, as is shown in Fig. 3.6, two searching steps are adopted to find the solution accurately. First of all, to prevent searching from trapping in a local solution, a rough searching with wider image lines, smaller grayscale changes, and relatively larger step size is employed, enabling to discover the areas that may contain the best solution quickly. Afterward, thinner image line features with larger grayscale changes, along with a smaller step size, are applied for a more precise calibration result. The switch between these two-step sizes and grayscale changes will happen when the proportion of the LiDAR points projected to the gray area exceeds a certain threshold.

For computation efficiency, an adaptive optimization method is proposed to enable the optimization toward the correct direction. In [8], they did calculations for 729 different values to compute the function score, which is an inefficient way as

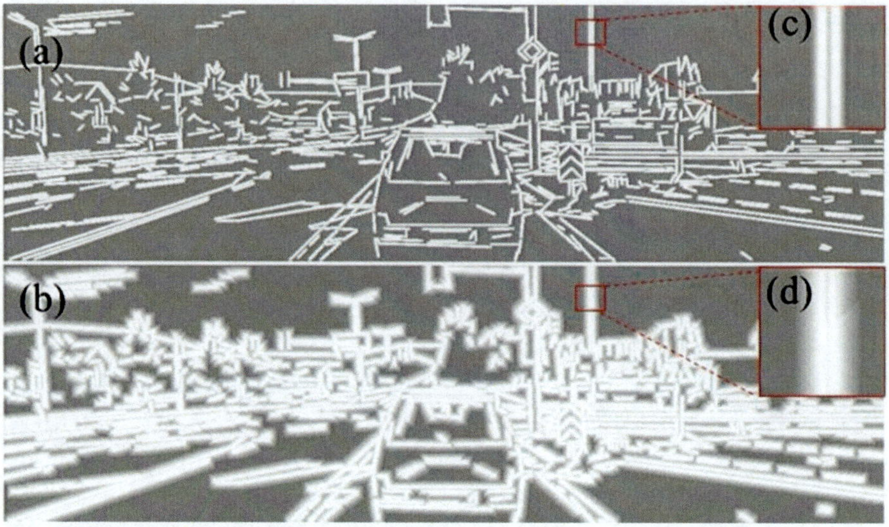

Fig. 3.6 Two different grayscale changes. (**a**) shows a larger grayscale change utilized in precise searching. (**b**) Shows a smaller grayscale change utilized in rough searching. Zoomed-in view (**c**) and (**d**) demonstrate the comparison of different grayscale changes more clearly

some steps are redundant. In this section, a searching method is applied to optimize the cost function. It will compare the current score with adjacent 728 scores. In this process, if the searching program finds parameters that have a higher score, it will stop the current searching process and begin a new searching process at the position providing a higher score. Besides that, this searching process will stop when reaching the set iteration count or finding the best score, thus being able to increase the computation efficiency. In addition, a sliding window is used to set the frames that the optimization progress should consider. In this section, three frames are utilized to prevent the optimization searching from a wrong direction or falling into a locally optimal solution. Therefore, the final optimized extrinsic parameters should exceed other parameters in all frames in the sliding window.

In conclusion, the two settings work together to get a robust and precise calibration result. The termination strategy enables a faster calibration. This process can be seen at Algorithm 1.

3.2.2 *Experiment*

In order to validate the proposed approach, experimental tests were conducted on KITTI dataset [4]. The KITTI dataset adopts different evaluation criteria for different tasks. For calibration tasks, our evaluation indicators are confidence and error (from displacement, pitch, yaw). During the experiment, we used Velodyne

Algorithm 1 Optimization process

Input: Image line features I_t, Horizontal F_h^t and vertical F_v^t line features at frame t, Initial extrainsic matrix T_t, last frame gray rate $gray\_rate$
Output: Calibrated extrainsic matrix;
Initialization: $score, max_{score} \leftarrow 0$.
if $gray\_rate > \gamma$ **then**
  $step\_size = \alpha_1$
else
  $step\_size = \alpha_2$
end if
Add disturbance based on step_size
for each $LiDAR\ point$ in F_h^t **do**
  $gray\_value = weight * T_t * p_t$
  $score+ = gray\_value$
end for
for each $LiDAR\ point$ in F_v^t **do**
  $gray\_value = (1 - weight) * T_t * p_t$
  $score+ = gray\_value$
end for
if $score > max\_score$ **then**
  $max\_score = score$
  Update current_parameters
end if
$gray\_rate = score/255/points\_num$
return current_parameters

HDL-64E LiDAR and a high-resolution color camera, and the scanning frequency of LiDAR was 10Hz. In the experiment, we simultaneously detect and correct the roll, pitch, and yaw deviations in real time and reflect the calibration standards by setting the confidence level. If the confidence in the three directions is greater than the given value, the calibration is considered accurate; if not, the confidence is low, that is, the calibration is considered inaccurate. We compared the calibration error with the ground truth and also tested the detection capability of error correction and the speed of deviation correction. The external parameters of the ground truth can be obtained from the calibration file, and the experimental data had been synchronized and corrected.

We carried out two experiments on different scenarios in the KITTI dataset. Scenario I gave the result in Fig. 3.7a and b, and Scenario II gave the result in Fig. 3.7c and d. In these two scenarios, we initially added a one-degree rotation bias on X-, Y-, and Z-axes and a 0.05-m transformation bias to the ground truth parameters. Then, a 0.5-degree rotation bias was added every ten frames. It should be pointed out that whether the one-degree rotation bias is positive or negative is randomized. During the experiment, we compared the calibration error with the ground truth. In addition, we tested the ability to detect miscalibration and the speed of correcting bias.

To evaluate the accuracy of the proposed approach, we computed the absolute value of the deviation of calibrated results from ground truth in two scenarios. The

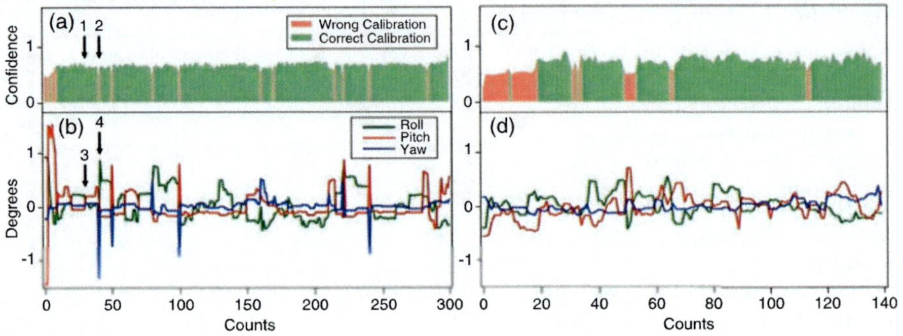

Fig. 3.7 Simultaneously detecting and correcting roll, pitch, and yaw deviations in real time. (**a**) shows the confidence of calibration results in Scenario I; (**b**) shows the calibration results of our method in Scenario I; (**c**) shows the confidence of calibration results in Scenario II; (**d**) shows the calibration results of our method in Scenario II. The green, red, and blue lines in (**b**) and (**d**) show the biases of roll, pitch, and yaw compared to ground truth. The red sections in (**a**) and (**c**) represent a low confidence of current parameters. On the contrary, the green sections in (**a**) and (**c**) show a high confidence of current calibration result

quantitative analysis of calibration deviation can be seen in Fig. 3.7. In Fig. 3.7b, the proposed method was able to simultaneously correct roll with a mean average error of 0.288° , pitch with a mean average error of 0.268°, and yaw with a mean average of 0.108° each. If the confidence below 0.7 was ignored, the error of roll, pitch, and yaw will decrease to 0.217, 0.209, and 0.082° . In Fig. 3.7d, the mean error of roll, pitch, and yaw are 0.217, 0.228, and 0.079°. If the confidence below 0.7 was ignored, the error of roll, pitch, and yaw will decrease to 0.159, 0.164, and 0.070°. Without counting the frames with manual error, the maximum error of roll, pitch, and yaw is always within 0.5° . The calibration results of yaw are the most precise since the LiDAR has a high horizontal resolution. Even though the LiDAR's resolution in the vertical direction is much lower and the 3D features in that direction are less frequently presented, the proposed approach can still achieve high accuracy due to the adaptive optimization algorithm and higher weight in this direction. Overall, an average rotational error of 0.12° across all dimensions is achieved, which is lower than most offline calibration techniques.

Additionally, we also tested the speed of correcting bias. Figure 3.7a and c represents the confidence of current calibration result. The extrinsic parameters were worse after artificially applying a bias, showing a larger angle deviation. In Fig. 3.7a and b, the black arrows 1, 2, 3, and 4 pointed at the frames of adding biases. For arrows 1 and 3 (30th frame), the proposed method corrected the bias immediately without showing the decline of confidence, while for the black arrows 2 and 4 (40th frame), the red section turned green within two frames (Fig. 3.7a) which means the bias was corrected within two frames. The same situation appeared in Scenario II (Fig. 3.7c and d). The overall calibration results in more scenarios on the KITTI dataset can be seen in Fig. 3.8, which demonstrates that the proposed method is applicable to different scenarios.

Fig. 3.8 The calibration results on several scenarios. Green points represent the projected LiDAR line points

3.3 Challenges and Prospect

Automatic extrinsic calibration of LiDAR and camera has very broad prospects, because accurate external parameters are essential for providing accurate information for the sensing system. This section proposes a point cloud line extraction method for extracting straight line features from the point cloud and introduces an adaptive optimization method and a result confidence estimation method to make the optimization procedure in the right direction. Of course, in the future, we also hope that the extracted line features can be more accurate and can have smaller deviations in scenes with complex or fewer features.

Currently, camera and LiDAR self-calibration algorithms for autonomous driving are facing some challenges. Our automatic calibration algorithm uses the original camera image and LiDAR point cloud for calibration, which has a high tolerance for data quality. We propose a method for automatic external calibration of LiDAR and camera based on line features and prove that line features can be extracted from point clouds and images with robust geometric features, which help to obtain accurate external parameters. Most of our experimental datasets contain simple and obvious line features, so they are easier to extract and identify. However, if there are a lot of pedestrians, vehicles, trees, and other debris on a very busy city street, the line features may not be easily extracted, resulting in an inaccurate calibration in this case. On the other hand, when it is dark or on a road in the field, the line features of the collected data are not very obvious. In this case, how to obtain the line features is also a challenge for us in the future.

3.4 Summary

With the growing importance of sensors used in the multimodal perception system in autonomous vehicles, the precise extrinsic parameters between these sensors are essential to provide a high-precision perception system. However, long-term operation under different loads can cause a slight change and drift from the extrinsic transformation matrix, which compromises perception accuracy. Therefore, it is crucial to automatically correct the miscalibration during operation.

In this chapter, we have proposed an online approach that can automatically calibrate the extrinsic parameters of LiDAR and camera. Different from previous automated methods, this new calibration approach requires no markers to be placed in the scene. And the proposed approach demonstrates that the line features from point clouds and images are robust features to correct calibration biases. The artificially added bias can be corrected within one or two frames, which is faster than other methods. In addition, we illustrate that the degree of confidence for the current calibration result can be computed and further utilized to improve the computation efficiency and accuracy.

For future work, we would like to evaluate the accuracy of extracted line features to reduce the calibration bias in scenes with few features. And the Monte Carlo method can be utilized to provide the initial parameters.

References

1. Castorena, J., Kamilov, U.S., Boufounos, P.T.: Autocalibration of lidar and optical cameras via edge alignment. In: 2016 IEEE International Conference on Acoustics, Speech and Signal Processing (ICASSP), pp. 2862–2866. IEEE, Piscataway (2016)
2. Fremont, V., Bonnifait, P., et al.: Extrinsic calibration between a multi-layer lidar and a camera. In: 2008 IEEE International Conference on Multisensor Fusion and Integration for Intelligent Systems, pp. 214–219. IEEE, Piscataway (2008)
3. Geiger, A., Moosmann, F., Car, Ö., Schuster, B.: Automatic camera and range sensor calibration using a single shot. In: 2012 IEEE International Conference on Robotics and Automation, pp. 3936–3943. IEEE, Piscataway (2012)
4. Geiger, A., Lenz, P., Stiller, C., Urtasun, R.: Vision meets robotics: the kitti dataset. Int. J. Robot. Res. **32**(11), 1231–1237 (2013)
5. Hess, W., Kohler, D., Rapp, H., Andor, D.: Real-time loop closure in 2d lidar slam. In: 2016 IEEE International Conference on Robotics and Automation (ICRA), pp. 1271–1278. IEEE, Piscataway (2016)
6. Kato, S., Takeuchi, E., Ishiguro, Y., Ninomiya, Y., Takeda, K., Hamada, T.: An open approach to autonomous vehicles. IEEE Micro **35**(6), 60–68 (2015)
7. Krizhevsky, A., Sutskever, I., Hinton, G.E.: Imagenet classification with deep convolutional neural networks. In: Advances in Neural Information Processing Systems, pp. 1097–1105 (2012)
8. Levinson, J., Thrun, S.: Automatic online calibration of cameras and lasers. In: Robotics: Science and Systems, vol. 2, p. 7 (2013)
9. Montemerlo, M., Becker, J., Bhat, S., Dahlkamp, H., Dolgov, D., Ettinger, S., Haehnel, D., Hilden, T., Hoffmann, G., Huhnke, B., et al.: Junior: the stanford entry in the urban challenge. J. Field Robot. **25**(9), 569–597 (2008)
10. Pandey, G., McBride, J.R., Savarese, S., Eustice, R.M.: Automatic targetless extrinsic calibration of a 3d lidar and camera by maximizing mutual information. In: AAAI. Citeseer (2012)
11. Qi, C.R., Su, H., Mo, K., Guibas, L.J.: Pointnet: Deep learning on point sets for 3d classification and segmentation. In: Proceedings of the IEEE Conference on Computer Vision and Pattern Recognition, pp. 652–660 (2017)

12. Ronneberger, O., Fischer, P., Brox, T.: U-net: Convolutional networks for biomedical image segmentation. In: International Conference on Medical Image Computing and Computer-Assisted Intervention, pp. 234–241. Springer, Berlin (2015)
13. Urmson, C., Anhalt, J., Bagnell, D., Baker, C., Bittner, R., Clark, M., Dolan, J., Duggins, D., Galatali, T., Geyer, C., et al.: Autonomous driving in urban environments: Boss and the urban challenge. J. Field Robot. **25**(8), 425–466 (2008)
14. Von Gioi, R.G., Jakubowicz, J., Morel, J.M., Randall, G.: LSD: a line segment detector. Image Process. Line **2**, 35–55 (2012)
15. Zhang, J., Singh, S.: Visual-lidar odometry and mapping: Low-drift, robust, and fast. In: 2015 IEEE International Conference on Robotics and Automation (ICRA), pp. 2174–2181. IEEE, Piscataway (2015)

Chapter 4
Multi-Sensor Object Detection

Abstract 3D object detection is becoming an indispensable functional module for environmental perception in autonomous driving, and LiDAR-based detection methods have made remarkable progress in terms of accuracy. However, point clouds often fail to distinguish objects with similar structures, leading to false detection. Therefore, other sensors and LiDAR fusion are naturally considered a solution. Nevertheless, current fusion methods are either limited to poor precision or efficiency. To this end, this chapter proposes a plug-and-play module named RI-Fusion to achieve the effective fusion of LiDAR and camera, and the module can be easily accessed by existing LiDAR-based algorithms. Furthermore, a particular fusion method of RaDAR and 16-line LiDAR based on multimodal and multi-scale fusion is proposed, called M^2-Fusion. The interaction is achieved by learning the features of each modality by exchanging the information of the intermediate feature layers with a self-attention mechanism. Experiments show that the method has better environmental adaptability and low cost.

4.1 Introduction

Recently, there are many studies using single-mode sensors such as cameras, LiDAR, and RaDAR for target detection in the past decades. However, each sensor has different flaws and drawbacks. The camera fails to obtain the actual distance of objects and is sensitive to illumination variation. These uncertain factors in autonomous driving, which is safety-critical, are unacceptable. The LiDAR can well compensate for these defects of cameras, and the LiDAR point cloud can adequately reflect real coordinates' information of the surrounding environment which is not subject to the change of illumination. Nevertheless, there are also numerous drawbacks of LiDAR. For instance, it is severely affected in rain and snow scenarios, and the point cloud lacks rich texture and color information as it exists in RGB images, which leads to the fact that different objects appear to have very similar structures in point cloud representation. RaDAR can detect longer distances without being affected by bad weather. However, 3D RaDAR sensor only has the horizontal position and velocity information. It lacks height information,

X. Zhang et al., *Multi-sensor Fusion for Autonomous Driving*,
https://doi.org/10.1007/978-981-99-3280-1_4

and its point cloud is such sparse that could not take over the detection task independently. 4D RaDAR captures three-dimensional position information and velocity information. Meanwhile, 4D RaDAR is insensitive to nonmetallic objects, and the data is also very sparse, which affects the recall rate of object detection.

To fully exploit the advantages among different sensors and fill gaps, we determine the detection method by fusing different data generated by multiple sensors. Compared with a single modality, multi-sensor fusion perception captures surrounding information through multiple sensors and receives more informative features. With a particular detection network, rich features with strong semantics and detail expression capability would lead to better detection performance. In this chapter, we elaborate on two multi-sensor fusion perceptions as LiDAR-image fusion and RaDAR-LiDAR fusion.

4.2 LiDAR-Image Fusion Object Detection

Cameras and LiDAR sensors are widely used in autonomous driving vehicles, but their limitations are apparent. The camera cannot obtain the distance information of the object and is very sensitive to the change of illumination. LiDAR can make up for these defects of the camera. The LiDAR point cloud can fully reflect the factual coordinate information of the surrounding environment and is not affected by the change of illumination. Therefore, the multimodal fusion of image and LiDAR features to complement their defects is a promising method.

There are still many datasets for 3D object detection. Due to space limitations, the author will not introduce them one by one. Instead, readers can choose a dataset suitable for their subject according to their research needs.

Therefore, the multimodal fusion of image and LiDAR features to complement their defects is a promising method; converting heterogeneous modalities into suitable representations is rewarding for fusion. As LiDAR point cloud can be converted to a compact 2D range image which makes it efficient to process, we propose a novel range-image fusion network to fuse point cloud and RGB image. The range image is the native representation and retains all the original information of the point cloud. This method decreases the difference in data representation between the point cloud and the RGB image. The position relationship between points in the point cloud is shown by the relative position between pixels in the range image. And the range image can be directly encoded using a CNN (convolutional neural network), which is effective for integrating RGB features. The main contributions of our work are as follows:

- We propose a plug-and-play module named RI-Fusion for the effective fusion of LiDAR and camera, and the module can be easily accessed to existing LiDAR-based methods.
- We propose a novel data fusion scheme for LiDAR point clouds and RGB images. By converting a point cloud to a range view, the semantic gap between

the point cloud and image is crossed, and feature fusion with RGB images is performed to enhance the accuracy of 3D object detection.
- Based on the self-attention mechanism, we propose a RI-Attention network to effectively combine the range image and RGB image and achieve remarkable improvements, especially on small objects such as pedestrians and cyclists.

4.2.1 RI-Fusion Framework

To effectively fuse the LiDAR point cloud and camera image, we propose a novel RI-Fusion framework to effectively fuse LiDAR point clouds and camera images, which can achieve remarkable improvements in accuracy. The overall framework adopts point clouds and RGB images as inputs and produces point clouds with fusion features that can be fed to common LiDAR-based methods to output predictions using 3D bounding boxes, as illustrated in Fig. 4.1. The core RI-Fusion module in this framework can be divided into three components. (1) Data preprocessing converts point cloud and RGB image inputs to the appropriate format for the fusion network. (2) An RI-Attention network is proposed to fuse range and RGB image features. (3) Point cloud recovery converts fusion features into point clouds.

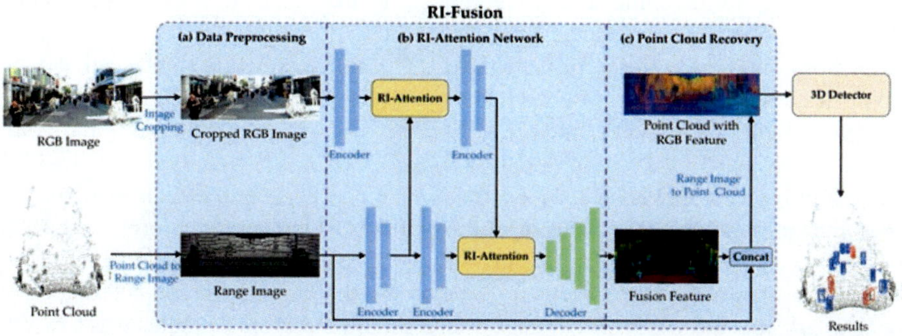

Fig. 4.1 Illustration of the RI-Fusion framework. In general, the RI-Fusion module can be plugged in front of the LiDAR-based 3D object detection framework. The fusion process of RI-Fusion can be divided into three parts: (**a**) Data preprocessing. The input point cloud is converted to a range image and the RGB image is cropped to obtain a suitable size. (**b**) RI-Attention network. The range image and cropped RGB image are performed by encoders, RI-Attention modules, and decoders in turn to achieve feature fusion. More details are given in Sect. 3.2. (**c**) Point cloud recovery. The fusion feature is concatenated with the range image to retain the spatial information which is utilized to recover the point cloud again. Finally, the point cloud with RGB features is fed into a 3D object detector for detection

4.2.1.1 Data Preprocessing

Point clouds were converted into range view representations to denote information more effectively and regularly. Unlike other projection methods, range views retain all original information and produce less loss in actual autonomous driving scenarios. In addition, range views converted using this approach are a dense representation of LiDAR, with features similar to RGB images. As a result, depth features can be quickly learned using a two-dimensional convolution layer. Thus, the converted distance image is more conducive to feature fusion with RGB images. The original LiDAR point cloud is assumed to be a matrix of size (n, m), where n represents the number of points in the scene and m indicates the features composed by each point. The KITTI dataset served as an example ($m = 4$), including the spatial coordinates X, Y, and Z and the LiDAR reflection intensity. Since point clouds are typically sparse and unstructured in space, conventional CNNs cannot be directly utilized for feature extraction, and range views are more conducive to extracting features from point clouds. The conversion of a LiDAR point cloud to a range view can be represented as:

$$\begin{bmatrix} u \\ v \end{bmatrix} = \begin{bmatrix} \frac{1}{2}\left[1 - \arctan(y, x)\,\pi^{-1}\right] \times w \\ \left[1 - (\arcsin(z, r) + f_{down}) \times f^{-1}\right] \times h \end{bmatrix}, \tag{4.1}$$

where x, y, and z are the spatial coordinates of each point, (u, v) is a pixel coordinate corresponding to the point in the range image, w and h are the width and height of the range image, r is the distance from each point to the origin, and $f = f_{up} + f_{down}$ is the horizontal pitch angle for the LiDAR. The above equation suggests that each point in the point cloud can be encoded with five-dimensional features, including distance, spatial coordinates, and reflection intensity. As a result, the original point cloud is converted into a range image of size $(5, w, h)$. A Velodyne 64E LiDAR was employed for the KITTI dataset used in the experiment, which involved 64 scan lines and a horizontal scan resolution of 0.2°. This approach produced 1800 points when scanning a circle, yielding a range image size of (5, 64, 1800).

Only a forward view of 90° is labeled in the KITTI 3D object detection benchmark, and the width of the range image should therefore be reduced as a result. Widths of 512, 768, and 1024 were adopted in these experiments, and a 2D CNN was efficiently utilized for feature extraction after the point cloud was converted into a 2D range view. Furthermore, because the resolution of laser RaDAR and RGB sensors is not consistent, the size of RGB images is larger than that of range images. As a result, there is no point cloud information corresponding to a wide area in the RGB image. However, it is necessary for these sizes to agree in the process of LiDAR and RGB information fusion, which inevitably leads to redundant extraction of RGB features. The removal of these redundant features can lead to significant information loss, especially in autonomous driving applications, which can degrade the high recall achieved using LiDAR and camera images. As such, we propose a new fusion scheme based on adaptive pooling to significantly boost the algorithm without losing essential information, which proved to be both effective and practical in the validation experiments. In addition, we utilized adaptive pooling instead of direct cropping to reduce information loss when adjusting the sizes of RGB images and range images. A series of comparative experiments were conducted using different sizes in the latter stages, to illustrate the influence of clipping size on RGB images.

4.2.1.2 RI-Attention Network

In the previous section, corresponding range view representations were acquired from point clouds, and the input range view was denoted as R. The KITTI dataset also provides RGB images for each point cloud, represented by I. In this study, R and I were input to two similar 2D convolution networks used to extract point clouds and RGB image features. It is worth noting that some pixels will inevitably overlap or exhibit holes during the conversion of the point cloud into R, due to quantization errors. For this reason, a dilated convolution was adopted in the encoder to produce a larger receptive field that differed from the common 2D convolution used for I. Each layer of the encoder module utilized a pooling layer to perform down-sampling and extract features from different receptive fields. RGB image features were then converted to the size of range features using adaptive pooling. Finally, range features (R_f) and image features (I_f) of the same size were produced and input to the range-image attention (RI-Attention) module for feature fusion. The resulting fused output for R_f and I_f is referred to as the RI feature (RI_f). This same process can be repeated using an RI-Attention module to extract multi-scale features. R_f perpetually extracted high-semantic features through dilated convolutions and a pooling layer, used to down-sample the results. RI_f also involved down-sampling after feature extraction, and the two features were then fed into another RI-Attention module. Since the output size was the same as that of the input after implementing RI-Attention, a fusion module could be added to each position in the network to achieve fusion of range and RGB features at different scales.

The proposed RI-Attention, a core self-attention module used to effectively fuse R_f and I_f, was repeated several times in the network. It is worth noting that the feature map was reshaped from $H \times W \times C$ to $(HW) \times C$ during self-attention, where $((HW))$ is the characteristic graph and C is the number of channels. Here, we compare the spatial features of RGB images with the feature maps of distance images, consider the similarity between the two modes, and then fuse RGB features conducive to object detection with distance features. This process is illustrated in Fig. 4.2, where the Q, K, and V matrices included in the attention mechanism denote query, key, and value, respectively. These features were obtained from the original input using a Conv1D layer represented as follows:

$$\begin{cases} Q = R_f \times W_q \\ K = I_f \times W_k \\ V = I_f \times W_v, \end{cases} \tag{4.2}$$

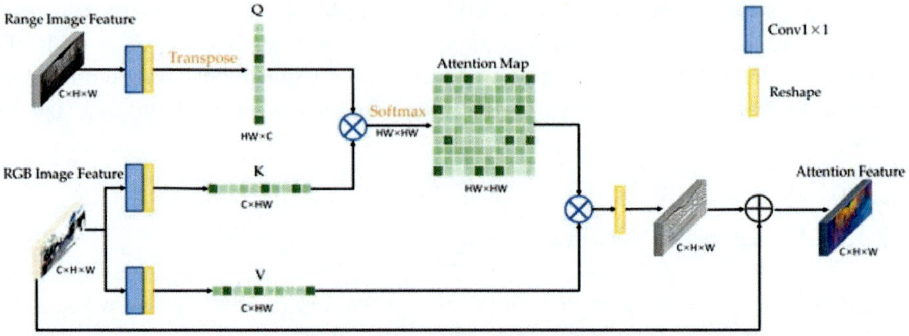

Fig. 4.2 Illustration of the RI-Attention module. The input range and RGB image feature are transformed and reshaped to $Q^{'}$, $K^{'}$, and $V^{'}$, which are all one-dimensional tensors. A multiplication is employed to establish a connection between range and RGB features. The relevant information is recorded in the attention map. Next, the tensor $V^{'}$ containing the RGB feature fuses the range information by multiplying it with the attention map. Finally, the original RGB feature is appended with the fusion feature to retain more RGB information

where W_q, W_k, W_v are weight matrices for linear layers that do not change the dimensions of tensors. The matrix product of Q and K produces an attention map used to establish the correlation of each element on a global scale. The matrices Q, K, and V are then transferred to three one-dimensional tensors: $Q^{'}$, $K^{'}$, and $V^{'}$. The product of $Q^{'}$ and $K^{'}$ yields the attention map M in a process that can be summarized as:

$$\begin{cases} Q^{'} = reshape(Q) \\ K^{'} = reshape(K) \\ M = Q^{'} \cdot K^{'T} \end{cases} \tag{4.3}$$

The next step requires normalizing the attention map and activating salient features in a nonlinear layer. A softmax function was then adopted as the activation to produce a normalized attention map $M^{'}$:

$$M^{'} = softmax(M) \tag{4.4}$$

Finally, the dot product of $V^{'}$ and $M^{'}$ provides selective fusion of RGB features through the guidance of a point cloud. Similarly, RGB image features are retained by concatenating I_f with fusion features. The resulting attention feature A_f can be expressed as:

$$A_f = M^{'} \cdot V^{'} + I_f \tag{4.5}$$

Since both RGB and range images comprise a frontal view, objects in individual scenes often appear small in the distance and large in the foreground. Feature pyramid networks (FPNs) are typically adopted in 2D object detection tasks and are often used to extract features at different scales. A similar approach was used for feature fusion in this study, which was conducive to the fusion of I_f at different feature levels. After two-layer feature fusion, the size of RI_f was equivalent to down-sampling the original input by a factor of 4. Fusion features were then input to the decoder network, and the up-sampling structure formed by the deconvolution network was used to restore feature maps to their original input size, allowing point cloud and RGB fusion features to be extracted from each point.

4.2.1.3 Point Cloud Recovery

The original range image was concatenated with fusion features to retain initial point cloud information. Features were then restored to a spatial point cloud representation based on depth and angle data from each pixel. This step can be expressed as:

$$p = \begin{bmatrix} x \\ y \\ z \end{bmatrix} = \begin{bmatrix} rcos\phi cos\theta \\ rcos\phi sin\theta \\ rsin\phi \end{bmatrix}, \tag{4.6}$$

where $p(x, y, z)$ is a coordinate in the point cloud, r is the depth value, and θ and ϕ are horizontal and vertical angles, respectively. The recovered point cloud with RGB features can then be fed to the downstream 3D object detection task.

4.2.2 *Experiment*

A set of experiments were conducted to demonstrate the remarkable improvements in 3D object detection achieved by the proposed RI-Fusion technique. In this section, we introduce the dataset used in the experiments and describe experimental details.

4.2.2.1 Dataset and Evaluation Metrics

The proposed method was evaluated using the KITTI 3D object detection dataset, a common benchmark used for autonomous driving studies that includes spatiotemporal registration data from cameras and LiDAR. The set includes 7481 training samples and 7518 test samples. Primary detection categories include car, pedestrian, and cyclist. These are divided into three difficulty levels (i.e., easy, moderate, and hard) depending on the object size, visibility, and truncation. A standard split methodwas used to divide the data into 3712 samples for training and 3769 samples for validation. Average precision, the official evaluation criterion for the KITTI 3D object detection benchmark, was also used as an evaluation metric in this study. In addition, a new technique was applied to calculate AP using 40 selected recall points, rather than the conventional value of 11. In this way, more accurate AP indexes were obtained, which provided a new standard for comparing our method with mainstream methods.

4.2.2.2 Implementation Details

The proposed network is a universal framework for the fusion of images and point clouds. The core RI-Fusion module in the network consists of three parts: data preprocessing, an RI-Attention network, and point cloud recovery. In the first step, 2/3 of the lower RGB image pixels are cropped and provided as input. The input point cloud is then converted to a range view representation of size (5, 64, 768), Eq. (4.1).

In the second step, RGB semantic features are acquired using a convolutional layer. The RGB feature map channel is then adjusted from 3 to 16 during the encoding process. In addition, adaptive pooling was utilized to modify the sizes of feature maps. Range images were also encoded using a convolution, and the feature map channel was adjusted from 5 to 16. The two resulting features each exhibited a size of (16, 32, 384). Subsequently, RGB and range image features were both input to the RI-Attention module, and features were then transformed to query, key, and value tensors using the method described in Section 3.2. The sizes of these tensors in the range and image feature maps were each (32, 32, 384) after three 1×1 convolution layers. The query and key tensors were reshaped into two one-dimensional tensors of size (32, 12, 288), and the attention map of size (12,288, 12,288) was calculated using matrix multiplication. The value tensor for the RGB image features was then multiplied by the attention map to perform attention-based fusion. Finally, the output was reshaped to a size of (32, 32, 384) and concatenated with RGB image features to produce attention features. In these experiments, two RI-Attention modules were added in the receptive field with scales of (32, 384) and (16, 192), yielding an output size of (32, 16, 192). A two-layer deconvolution sampling module was then adopted to recover the input size and produce a final output size of (8, 64, 768).

In the last step, fusion features are recovered from the point cloud. In addition, the original range image with an intensity value and spatial coordinates of x, y, and z was concatenated with fusion features to produce an output size of (12, 64, 768). Features were also recovered to point clouds using Eq. (4.6). Each point consisted of 12-dimensional features including an 8-dimensional RGB value acquired using the fusion process discussed above. This fusion point cloud can then be provided to other mainstream LiDAR-based detectors.

The RI-Fusion module was implemented with PyTorch 1.6. The effectiveness of the proposed fusion scheme was evaluated by selecting promising 3D objects detected in the MMDetection3D codebase. The AdamOptimizer and cosine annealing were included as optimization strategies. All experiments were trained on eight NVIDIA RTX 2080Ti GPUs using an initial learning rate of 0.001 for each detector. Data augmentation strategies were widely adopted in the training process, including image mirroring around the X-axis and Y-axis, random rotation around the Z-axis through a range of $[0, \pi/4]$, and scaling of $[0.95, 1.05]$. The "copy paste" methodwas also included, since a wide range of backgrounds were present in the training set. This added ground truth objects to empty background scenes and increased the proportion of positive samples in the training set.

4.2.2.3 Results

The performance of the proposed RI-Fusion module was assessed using several comparative experiments involving state-of-the-art fusion methods applied to the KITTI benchmark. The results indicated that RI-Fusion achieved better accuracy for Part A^2, as shown in Table 4.1. Furthermore, several experiments were conducted to verify the effects of inserting RI-Fusion into existing detectors, including PointPillars, SECOND,and Part A^2. These experiments were based on the MMDetection3D framework, an open-source object detection toolbox developed with PyTorch. Comparative 3D object detection results are shown in Table 4.2, with a baseline established using model zoo in the MMDetection3D codebase. As shown in the tables, each of the results produced by PointPillars and SECOND improved with the addition of RI-Fusion, while most of the Part A^2 results improved despite a slight reduction in some cases. Remarkable improvements were also observed in mAP, especially for the pedestrian and cyclist classes. PointPillars with RI-Fusion achieved an overall mAP increase of 3.61 for moderate levels. Similarly, the yield was +7.59% for moderate levels and +8.17% for hard levels in the pedestrian class. The cyclist category showed similar improvements. Comparative results for BEV object detection are provided in Table 4.3 and are consistent with 3D object detection. These results demonstrate that the proposed plug-and-play RI-Fusion module is conducive to enhancing the overall performance of LiDAR-based detectors.

Table 4.1 Performance comparison of our method with previous fusion methods on the KITTI validation 3D detection benchmark

	3D mAP(Car)		
Methods	Easy	Mod.	Hard
MV3D	71.29	62.68	56.56
ContFuse	82.54	66.22	64.04
F-PointNet	83.76	70.92	63.65
AVOD-FPN	86.42	77.10	76.11
PI-RCNN	88.27	78.53	77.75
Part A2 + RI-Fusion (ours)	**89.42**	**79.07**	**78.41**

Bold values highlight the significant improvement in performance

Table 4.2 Comparative results on the KITTI validation 3D detection benchmark. All methods with RI-Fusion module show the improvements in 3D overall mAP on the moderate split. Although there is a slight decline in some categories, most of the performance has been improved, especially for pedestrian and cyclist. The modalities are LiDAR(L) and images(I)

		mAP	Car			Pedestrian			Cyclist		
Method	Modality	Mod.	Easy	Mod.	Hard	Easy	Mod.	Hard	Easy	Mod.	Hard
PointPillars	L	59.53	82.76	74.92	67.86	49.96	44.50	39.32	77.89	59.19	56.63
+RI-Fusion	L and I	63.14	85.62	75.35	68.31	57.44	52.09	47.49	81.44	61.97	59.56
Delta	Δ I	+3.61	+2.86	+0.43	+0.45	**+7.48**	**+7.59**	**+8.17**	**+3.55**	+2.78	+2.93
SECOND	L	64.41	87.50	76.91	73.88	58.56	51.46	47.06	79.56	64.85	61.23
+RI-Fusion	L and I	67.39	87.59	77.00	74.61	63.06	57.15	52.16	81.01	68.01	64.68
Delta	Δ I	+2.98	+0.09	+0.08	+0.72	**+4.50**	**+5.69**	**+5.10**	+1.45	+3.15	+3.45
Part A^2	L	67.86	89.40	79.01	78.46	62.24	55.84	49.94	89.67	68.73	65.63
+RI-Fusion	L and I	69.13	89.42	79.07	78.41	62.00	55.93	50.37	88.83	72.39	67.92
Delta	Δ I	+1.27	+0.02	+0.06	−0.05	−0.24	+0.09	+0.43	−0.84	**+3.66**	+2.29

Bold values highlight the significant improvement in performance

Table 4.3 Comparative results on the KITTI validation BEV detection benchmark. All methods with RI-Fusion module show an improvement in BEV overall mAP on the moderate split. The performance of pedestrian and cyclist categories has been improved significantly on easy, moderate, and hard split, and there is a slight decrease in the car category. The modalities are LiDAR(L) and images(I)

		mAP	Car			Pedestrian			Cyclist		
Method	Modality	Mod.	Easy	Mod.	Hard	Easy	Mod.	Hard	Easy	Mod.	Hard
PointPillars	L	66.97	90.14	85.26	79.81	57.80	52.53	47.50	79.96	63.10	59.34
+RI-Fusion	L and I	69.27	89.62	86.30	79.60	63.15	57.26	52.61	82.53	64.25	61.68
Delta	Δ I	+2.30	-0.52	+1.04	−0.21	**+5.35**	**+4.73**	**+5.11**	+2.57	+1.15	+2.34
SECOND	L	72.08	89.29	87.12	84.92	65.59	60.20	53.37	82.55	68.93	64.82
+RI-Fusion	L and I	73.48	89.68	86.38	85.36	68.37	61.91	55.18	85.66	72.16	68.70
Delta	Δ I	+1.40	+0.40	−0.74	+0.44	+2.78	+1.71	+1.81	+3.12	+3.23	**+3.88**
Part A^2	L	72.44	90.24	88.02	87.25	63.99	57.77	54.48	90.07	71.55	68.21
+RI-Fusion	L and I	73.43	90.28	87.92	86.12	65.30	58.69	55.21	89.54	79.06	76.21
Delta	Δ I	+0.99	+0.05	−0.10	−1.13	+1.31	+0.93	+0.74	−0.53	**+7.52**	**+8.00**

Bold values highlight the significant improvement in performance

Runtime is another essential indicator used for evaluating 3D object detection results, especially for autonomous vehicles. As such, the inference time was evaluated for our proposed technique running on a single NVIDIA RTX 2080 Ti GPU, as shown in Fig. 4.3. The reasoning time was relatively short (0.00898 s) since most network operations involved matrix multiplications that are accelerated in CUDA. In addition, results showed that the addition of RI-Fusion to the three baseline methods significantly improved overall mAP values at moderate levels, despite slight increases in inference times. In addition, the running frame rate for PointPillars with RI-Fusion remained above 26.1 Hz. Some qualitative results are

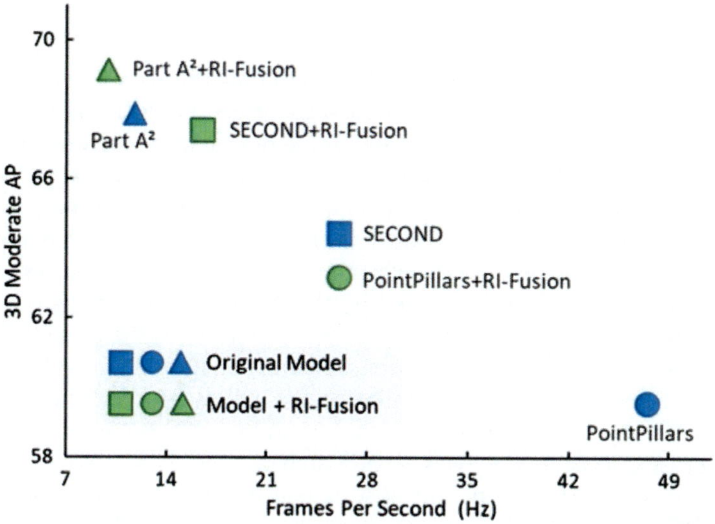

Fig. 4.3 Comparative results on the accuracy and inference time between the baseline models and our proposed method

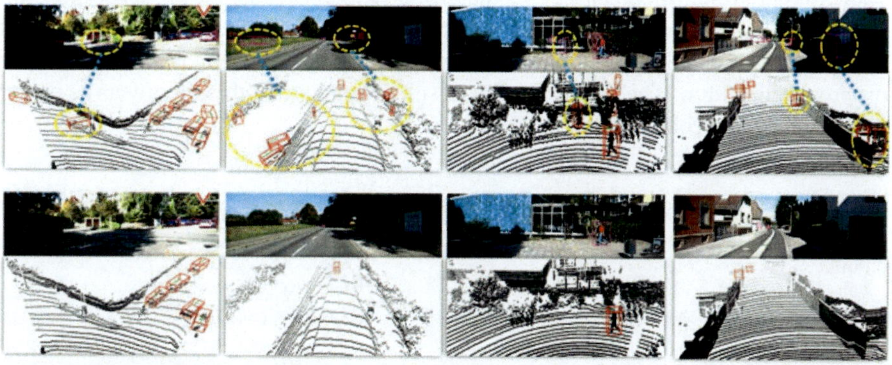

Fig. 4.4 Qualitative results of our RI-Fusion method (down) on the KITTI dataset compared with Part A^2 (top). The yellow circles represent incorrect results in Part A^2, which are corrected by our RI-Fusion method. Most of these false detections are caused by similar-shaped structures

also provided in Fig. 4.4, to demonstrate the advantages of our proposed technique. Specifically, detection results with Part A^2 serving as the baseline are shown in the first row of the figure. Results produced using our method are shown in the second row. It is evident from the figure that multiple incorrect detections occurred in Part A^2, primarily due to weak detection in point clouds with similar spatial structures. However, this effect was reduced significantly using our method. In addition, the introduction of RGB images effectively enhanced the differentiation of objects with similar point clouds and increased detection accuracy for small objects such as pedestrians and cyclists.

4.2.2.4 Ablation Studies

A series of ablation experiments, based on the PointPillarsframework, were conducted to evaluate the influence of various parameters on algorithm performance. Multiple sizes in the range images caused some quantitative loss when converting between point clouds and range views. As such, range images of sizes (48, 512), (64, 768), and (64, 1024) were used in the ablation experiment to study the effects of differing scales. The results shown in Table 4.4 indicate that overall mAP increased with range scale. The RI-Fusion module utilized RGB images to extract features from attention maps acquired by the fusion of range images and RGB image features. Range images were also adopted for use in the same process. As such, comparative experiments were conducted to demonstrate the effects of each technique. The results are shown in Table 4.5 and indicate that both fusion methods improved 3D object detection, while the accuracy achieved with our preferred technique (using RGB images) was slightly higher.

Additional comparative experiments were conducted to determine the impact of different RGB image sizes on the accuracy of the fusion model, as discussed in Sect . 4.3. RGB images input to PointPillars (with the RI-Fusion module included) were cropped by 3/4, 2/3, 1/2, and 1/3 of the lower image portion. As shown in Tables 4.6 and 4.7, the accuracy of the cropped images was not reduced and even improved in some cases, indicating the upper region of the original image included noisy data. After consideration, 2/3 of the lower image region was input to our proposed algorithm. The corresponding results demonstrate the effectiveness of our proposed fusion technique.

Table 4.4 The effects of different range image scales on mAP

Size of range view	Epoch	mAP
PointPillars	80	59.53
+RI-Fusion (48, 512)	80	61.66
+RI-Fusion (64, 768)	80	62.73
+RI-Fusion (64, 1024)	80	**63.28**

Bold values highlight the significant improvement in performance

Table 4.5 Comparative experiments of the different fusion methods on the KITTI validation 3D detection benchmark

	mAP	3D mAP (Car)		
Methods	Mod.	Easy	Mod.	Hard
PointPillars	59.53	82.76	74.92	67.86
PointPillars (V from range image)	62.34	83.61	74.25	67.73
PointPillars (V from RGB, ours)	**63.14**	**85.62**	**75.35**	**68.31**

Bold values highlight the significant improvement in performance

Table 4.6 Comparative experiments of different crop sizes of RGB image on the KITTI validation 3D detection benchmark using on PointPillars with our RI-Fusion module

	Car			Pedestrian			Cyclist		
RGB image	Easy	Mod.	Hard	Easy	Mod.	Hard	Easy	Mod.	Hard
All	73.86	64.84	61.41	47.96	43.97	39.91	73.27	55.95	53.61
3/4	84.07	72.66	67.30	55.48	50.10	45.74	79.08	59.85	57.95
2/3	85.62	75.35	68.31	57.44	52.09	47.49	81.44	61.97	59.56
1/2	83.80	73.69	67.49	56.05	50.87	45.59	82.23	64.40	60.70
1/3	85.45	75.21	68.14	54.99	50.10	45.20	79.27	59.41	56.50

Table 4.7 Comparative experiments of different crop sizes of RGB image on the KITTI validation BEV detection benchmark using on PointPillars with our RI-Fusion module

	Car			Pedestrian			Cyclist		
RGB image	Easy	Mod.	Hard	Easy	Mod.	Hard	Easy	Mod.	Hard
All	88.27	83.52	78.70	55.35	51.37	46.40	74.87	58.27	55.63
3/4	89.59	86.20	79.39	61.90	55.76	51.08	85.02	65.56	62.46
2/3	89.62	86.30	79.60	63.15	57.26	52.61	82.53	64.25	61.68
1/2	89.76	85.58	79.38	62.36	56.10	51.49	84.01	68.06	64.90
1/3	90.09	86.37	79.68	60.41	54.98	48.95	81.19	63.80	60.82

4.3 RaDAR-LiDAR Fusion Object Detection

Multimodal fusion is a crucial method in 3D object detection for autonomous driving to overcome single sensor perception's inherent shortcomings. Unfortunately, most fusion methods focus on camera and high-line LiDAR sensors, but RaDAR and low-line LiDAR fusion are rarely considered, especially for 4D RaDAR, a solution with commercial prospects. Therefore, this chapter proposes a novel 4D RaDAR and 16-line LiDAR fusion method to fill this gap. One primary challenge is that the point cloud from these two sensors contains different features and distributions. Another challenge is that both point clouds appear low density. Therefore, it performs poorly when integrating two-point clouds directly. To solve these problems, we propose a particular fusion method of 4D RaDAR and 16-line LiDAR based on multimodal and multi-scale fusion, called M^2-Fusion. We utilize the self-attention mechanism to learn each modality's features and then exchange intermediate feature layer information to realize the interaction. Specific to the current fixed resolution problem of voxel division, we propose a novel center-based multi-scale fusion method to extract features in the original point cloud. Then, these features are connected with the RPN head to realize 3D object detection.

Furthermore, we present a data preprocessing method based on Gaussian distribution that effectively decreases data noise to reduce errors caused by point cloud divergence of 4D RaDAR data in the x-z plane. A series of experiments were conducted using the Astyx HiRes 2019 dataset, including the calibrated 4D RaDAR and 16-line LiDAR data, to evaluate the proposed fusion method. The results demonstrated that our fusion method compared favorably with state-of-the-art algorithms, producing mAP (mean average precision) increases of 5.64% and 13.57% for 3D and BEV (bird's-eye view) detection of the car class at a moderate level, respectively. The code and pre-trained models are available at https://github.com/adept-thu/M2-Fusion-new-.git.

RaDAR has been receiving increased attention from industrial and academic communities as a potential alternative. These systems transmit millimeter waves and record the returned beams. The object distance can also be determined by measuring the round-trip time and velocity using the Doppler effect. In addition, the strong penetrability of millimeter-wave signals allows for long-range observations (usually over 200 m) in inclement weather, such as heavy rain and fog. 3D RaDAR systems, typically employed as an auxiliary measurement for other sensors, collect the horizontal position and velocity information (x, y, v) using antennas positioned only in the x and y directions. Usually, the azimuth angle resolution is over 10°, and it has no vertical angle resolution. In contrast, 4D RaDAR sensors use antennas positioned in three directions to obtain 3D position and velocity information (x, y, z, v), thus providing denser point clouds. The azimuth angle resolution is about 1°, and the vertical angle resolution is about 2°. For instance, the Astyx HiRes 2019 [5] dataset used in this work includes more than 1000 4D RaDAR points per frame. Compared to 3D RaDAR, 4D RaDAR significantly improves angle resolution and vertical ranging capability, enabling denser point cloud acquisition.

High-level autonomous driving requires multimodal fusion [1, 2, 8, 9], to achieve sufficient and robust detection performance. Fused data are often collected from a variety of sensors including 3D RaDAR, cameras, LiDAR, and 4D RaDAR. Specifically, 4D RaDAR offers substantial long-range (detection range over 200 m) and robustness advantages, while LiDAR offers higher accuracy. A more practical and cost-effective 3D object detection method could thus be developed by fusing these modalities effectively. Moreover, there are few research works on the fusion of 4D RaDAR and LiDAR. However, 4D RaDAR data is sparser than LiDAR, making it difficult to estimate objects' shapes and sizes. Furthermore, the range and height of 4D RaDAR point clouds are fuzzy due to aliasing and false detection caused by clusters and multi-path echoes. Accordingly, some objects are highly visible in LiDAR (e.g., cars and pedestrians), but appear blurry in RaDAR. Thus, effectively combining these two types of point cloud features is a critical problem. One intuitive solution would be to directly splice the features extracted from the two modalities to achieve fusion. Nevertheless, this strategy suffers from a low correlation between LiDAR and RaDAR data since the spatial distributions vary significantly across point clouds for the two modalities. To date, inter-modality relationships between 4D RaDAR and LiDAR have rarely been investigated in a framework used for object detection.

Motivated by the above discussion, as far as the authors know, we are the first to propose a 3D object detection fusion method for 4D RaDAR and LiDAR, termed multimodal and multi-scale fusion (M^2-Fusion). This process consists of two core modules: interaction-based multimodal fusion (IMMF) and center-based multi-scale fusion (CMSF). Through a large number of comparative experiments

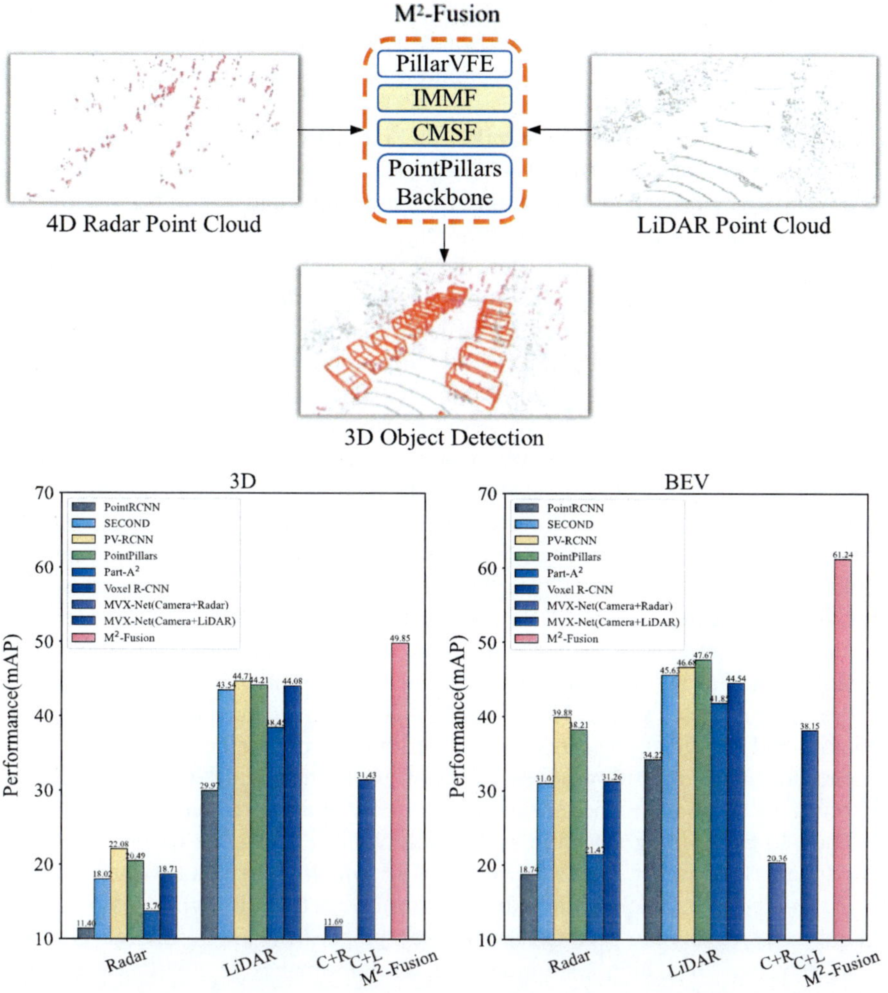

Fig. 4.5 The overall concept of M^2-Fusion processing both sparse 4D RaDAR and LiDAR point clouds simultaneously. We propose two novel modules termed IMMF and CMSF, which remarkably improve detection accuracy. The average precision bar charts above compare M^2-Fusion and seven mainstream object detection networks (PointRCNN [7], SECOND [10], PV-RCNN [6], PointPillars [4], and MVX-Net [8]). As shown, our method outperforms the others by a large margin in both 3D and BEV object detection. "C+R" and "C+L" mean "Camera + RaDAR" and "Camera + LiDAR," respectively

with existing mainstream methods, state-of-the-art 3D detection performance is achieved by significantly improving feature expression capabilities using a self-attention mechanism and multi-scale fusion (see Fig. 4.5).

One of the primary challenges in multimodal fusion is to maintain a strong relationship between different features of the same object from each modality.

The self-attention mechanism is a viable approach for accomplishing feature interaction, allowing the parameters from different modalities to learn from each other. Therefore, we propose an interaction-based multimodal fusion method termed IMMF. The module fuses and updates each 4D RaDAR and LiDAR feature using the attention-weighted information from the other modality. Then, the information flow is then transmitted to each modality to capture complex intra-modal relationships. As a result, the attention weights within each modality condition in the 4D RaDAR and LiDAR features are mutually reinforcing.

Secondly, high sparsity of 4D RaDAR and LiDAR data often necessitates the enhancement of feature expressions in each modality. For instance, voxel-based frameworks offer excellent performance and are considered the mainstream technology for 3D object detection [10]. In this process, a 3D space is divided into regular 3D voxels, and a convolutional neural network is utilized to extract features for compression into a bird's-eye view feature map for 3D object detection. However, due to the uneven density of point clouds in 3D spaces, setting a single scale for default voxel grids is insufficient to represent all information in a given scenario. Smaller grids produce more refined localization and features, while larger grids are faster as a result of fewer non-empty grids and smaller feature tensors. Multi-scale methods, which employ the benefits of both small and large grids, have shown great potential in various 3D object detection tasks [3]. One intuitive solution would be to directly merge voxel features from both high-resolution and low-resolution data as input to the next backbone, thus forming an aggregated feature across various voxel scales. However, this method inevitably causes tremendous memory and computational pressure on the limited onboard computing resources. Moreover, the superposition of feature maps at different scales does not highlight object characteristics and the fuzziness of 4D RaDAR may produce additional noise.

Thus, to better integrate multi-scale point cloud features, we propose a center-based multi-scale fusion method termed CMSF, which includes a two-step strategy. In the first step, key points are extracted from objects with higher scores. In the second step, voxels are selected around key points in a low-resolution grid and then converted to a high-resolution grid. Pseudo-images of all scales are then standardized using an adaptive max pooling operation.

In addition, we propose a data preprocessing method based on Gaussian normal distribution to reduce data noise of 4D RaDAR. Generally, 4D RaDAR sensors have good ranging accuracy, but insufficient resolution in the vertical direction leads to poor accuracy in the z-axis direction and introduces considerable data noise. We utilize the Gaussian normal distribution to obtain the statistic rule of 4D RaDAR point cloud distribution in the recent Astyx HiRes 2019 [5] dataset, a 3D object detection dataset that includes 4D RaDAR data. Then, noise in the 4D RaDAR point cloud is corrected to improve data quality.

The primary contributions of our work are as follows:

- We are the first to propose a novel fusion method termed M^2-Fusion that utilizes 4D RaDAR and LiDAR data for 3D object detection in autonomous driving, offering higher precision and especially remote detection accuracy.

- We introduce an interaction-based multimodal fusion method termed IMMF. This module can learn correlations between two modalities by utilizing attention mechanisms, emphasizing critical features, and reducing noise to improve accuracy and robustness for 3D object detection.
- We develop a center-based multi-scale fusion method termed CMSF, which combines the advantages of small-scale and large-scale features. We demonstrate that selecting voxels in a particular range around key points results in better efficiency and accuracy than other methods using only one quantization scale.
- We conduct extensive experiments using the Astyx HiRes 2019 dataset to evaluate M^2-Fusion. Experimental results showed that our method achieves state-of-the-art performance.

4.3.1 Preprocessing of 4D RaDAR Point Clouds

4D RaDAR is equipped with azimuthal and vertical angle antennas to measure horizontal and vertical positions. However, the number of vertical angle antennas of 4D RaDAR is often small, which will cause resolution issues. For example, due to aperture limitations, various objects with differing angles for the same speed and distance conditions may be challenging to distinguish in the azimuthal direction. Specifically, the theoretical vertical angle of the RaDAR in Astyx HiRes 2019 is θ_i degrees, but the actual one is higher than $9 \times \theta_i$ degrees, as shown in Fig. 4.6a. We observe that considerable 4D RaDAR points are below ground which affects the detection accuracy. Therefore, we propose a novel method for addressing noise in the data.

We use Gaussian normal distribution to determine if the vertical angle θ_t is in the normal range based on Shapiro-Wilk (S-W) test. Here, we focus on two descriptive statistics, i.e., skewness and kurtosis, which can help determine if the point cloud

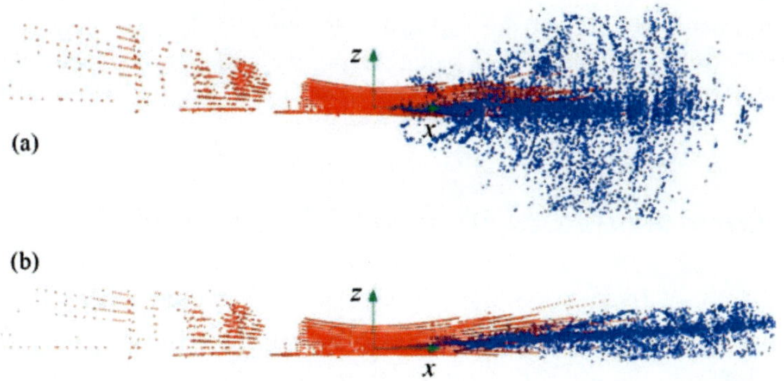

Fig. 4.6 *Y*-axis views of 4D RaDAR (blue) and LiDAR (red) point clouds, including (**a**) the original data and (**b**) the processed data

conforms to the Gaussian normal distribution. Specifically, the symmetry of the distribution measure, i.e., the inequality between the left and right distribution tails, can be verified using skewness value. The peakedness and heaviness of the distribution tails can be confirmed using kurtosis value. The skewness (g_1) and kurtosis (g_2) are shape parameters of the point cloud, which are calculated frame by frame as follows:

$$\begin{cases} g_1(t) = E[(\frac{\theta_t - \mu}{\sigma})^3] \\ g_2(t) = E[(\frac{\theta_t - \mu}{\sigma})^4], \end{cases} \tag{4.7}$$

where θ_t denotes the divergence angle of the point cloud in 4D RaDAR, μ is the mean value, σ is the standard deviation of divergence angle, and E refers to the mean operation.

As the maximum divergence angle for the 4D RaDAR is far beyond the sensor setting range, we reduce the point cloud elevation in the x-z plane so that all vertical slopes are limited to within θ_m. When the absolute value of kurtosis and skewness are both less than 1, we assume the data frame conforms to the Gaussian normal distribution, and then θ_m is set as the mean value of the vertical angle. Otherwise, if this frame does not conform to the Gaussian normal distribution, then we set θ_m as the median value. Finally, we update the coordinates by the following equation:

$$\begin{cases} x_t' = cos(\frac{2\theta_t}{\theta_m} arctan\frac{z_t}{x_t})\sqrt{x_t^2 + z_t^2} \\ z_t' = sin(\frac{2\theta_t}{\theta_m} arctan\frac{z_t}{x_t})\sqrt{x_t^2 + z_t^2}, \end{cases} \tag{4.8}$$

where x_t and z_t are original coordinates of the point cloud, x_t' and z_t' are the coordinates after preprocess, and θ_m is the angle at which preprocessed data are compressed.

After the above data preprocessing of 4D RaDAR point cloud, the noise data can be improved to conduct the following detection task. The intuitive effect is shown in Fig. 4.6b, which illustrates that the point cloud data is transformed within a reasonable range.

4.3.2 Interaction-Based Multimodal Fusion (IMMF)

A comparison of 4D RaDAR and LiDAR data suggests their forms are similar, though RaDAR exhibits weak reflection echoes from vehicles, pedestrians, and cyclists. However, the RaDAR point cloud for the background environment is also much more extensive than that of the recognition objects. Existing network structures are better at identifying particular objects in uniform backgrounds. Therefore,

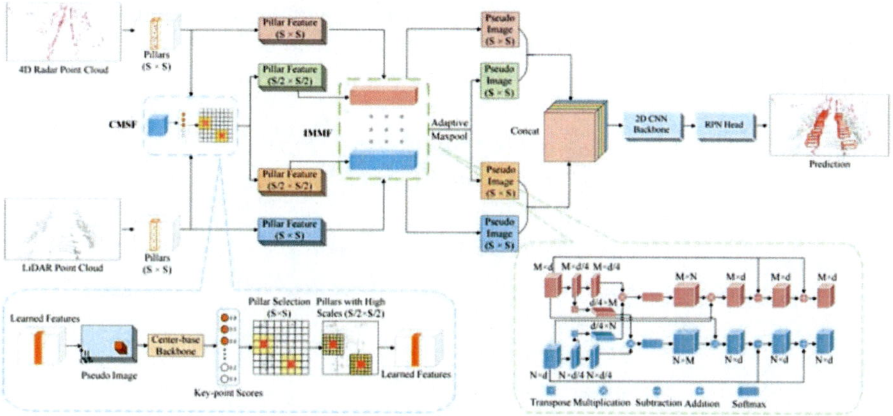

Fig. 4.7 The overall framework of our proposed M^2-Fusion method consisting of four sub-networks. (1) IMMF: the module in the green box demonstrates multimodal fusion, which learns correlations between two modalities by utilizing an attention mechanism. (2) CMSF: the module in the blue box performs multi-scale feature extraction by dividing selected pixels for key points into small-scale pixels. (3) 2D CNN backbone. (4) RPN (Region Proposal Network) head

it is essential to focus our limited resources on the core portion when integrating the two modalities, for which we propose an interaction-based multimodal fusion approach (IMMF). Precisely, the tensor decomposed from a feature in the attention mechanism (between modalities) constitutes a weight matrix that exerts network learning attention and exchanges inter-modality tensors. In other words, LiDAR representations guide the attention of 4D RaDAR, and 4D RaDAR representations direct the attention of LiDAR.

The network structure in the green dotted box in Fig. 4.7 shows the detailed IMMF module, a symmetrical structure that consists of two self-attention models. IMMF facilitates interaction between modalities by exchanging tensors in symmetric structures and guiding the network to learn more valuable features. Following PointPillars, we divide the input point cloud into M×N pillars, where points in each pillar are expanded from four dimensions (x, y, z, r) to ten dimensions $(x, y, z, r, x_c, y_c, z_c, x_p, y_p, z_p)$. The calculations are as follows:

$$\begin{cases} (x_c, y_c, z_c) = (x, y, z) - (x_m, y_m, z_m) \\ (x_p, y_p, z_p) = (x, y, z) - (x_g, y_g . z_g), \end{cases} \tag{4.9}$$

where r represents the reflectivity, (x_c, y_c, z_c) is the deviation of each point cloud relative to the pillar center, (x_m, y_m, z_m) is the center coordinate of each pillar, (x_p, y_p, z_p) is the deviation of each point cloud from the grid center, and (x_g, y_g, z_g) is the center coordinate of each grid.

We extract features of each pillar from LiDAR and 4D RaDAR point clouds and set F_L and F_R, the initial features for LiDAR and 4D RaDAR, as the input to the IMMF module. We then adopt convolutional networks to improve computational efficiency with the strategy that reduces the dimension of $F_{L_{64}}$ and $F_{R_{64}}$. The reduced feature dimensions are both 16, denoted by $F_{L_{16}}$ and $F_{R_{16}}$. The corresponding weight matrices F_{L_w} and F_{R_w} can then be calculated as follows:

$$\begin{cases} F_{L_{64}} = Maxpool(Linear(F_L)) \\ F_{R_{64}} = Maxpool(Linear(F_R)), \end{cases} \tag{4.10}$$

$$\begin{cases} F_{L_{16}} = Conv(Maxpool(Linear(F_L))) \\ F_{R_{16}} = Conv(Maxpool(Linear(F_R))), \end{cases} \tag{4.11}$$

$$\begin{cases} F_{L_w} = Softmax((F_{L_{16}})^T F_{R_{16}}) \\ F_{R_w} = Softmax((F_{R_{16}})^T F_{L_{16}}), \end{cases} \tag{4.12}$$

where $Conv(\cdot)$ denotes a convolutional layer, $Maxpool(\cdot)$ represents a maximum pooling layer, $Linear(\cdot)$ indicates a fully connected layer, and $Softmax(\cdot)$ is a softmax function.

The sizes of F_{L_w} and F_{R_w} are $M \times N$ and $N \times M$, respectively, where M is the number of LiDAR features and N is the number of RaDAR features. These two weight matrices contain modal information for each other. We then multiply the F_{L_w} term on the right by $F_{R_{64}}$. The resulting feature size is $M \times d$, since F_{L_w} has a size of $M \times N$ and $F_{L_{64}}$ has a size of $N \times d$ (d is the characteristic dimension). This feature is consistent with the size of $F_{L_{64}}$ and is subtracted from and then added to $F_{L_{64}}$ after the linear layer, normalization layer, and activation function, to acquire interactive LiDAR features F_{L_t}. Similarly, the same operation is conducted to obtain interactive RaDAR features F_{R_t}. This process can be represented as:

$$\begin{cases} F_{L_m} = ReLU(BN(LN(F_{L_w} F_{R_{64}} - F_{L_{64}}))) \\ F_{R_m} = ReLU(BN(LN(F_{R_w} F_{L_{64}} - F_{R_{64}}))), \end{cases} \tag{4.13}$$

$$\begin{cases} F_{L_t} = F_{L_m} + F_{L_{64}} \\ F_{R_t} = F_{R_m} + F_{R_{64}}, \end{cases} \tag{4.14}$$

where $ReLU(\cdot)$ is a rectified linear unit activation function, $BN(\cdot)$ denotes a batch normalization function, F_{L_m} and F_{R_m} are intermediate variables, and $LN(\cdot)$ represents a linear layer.

4.3.3 Center-Based Multi-Scale Fusion (CMSF)

The division method used in PointPillarscan divide point clouds into high-resolution or low-resolution pillars, depending on the voxel size. High-resolution voxel divisions can learn more features and improve the resulting detection accuracy

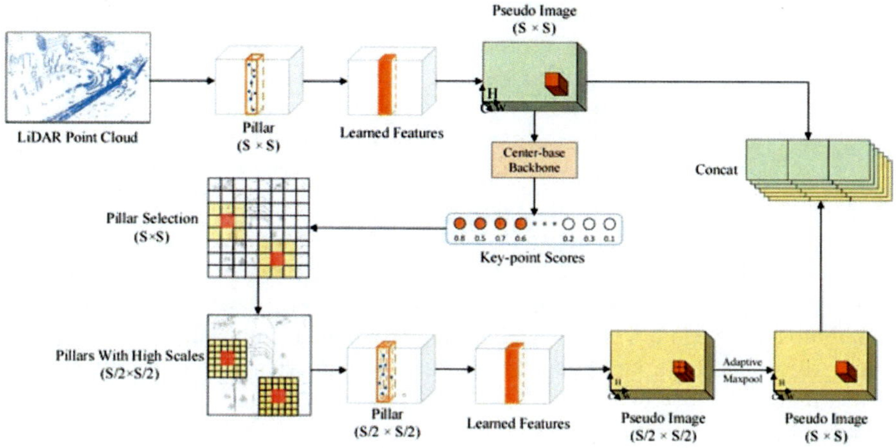

Fig. 4.8 The network framework for the CMSF module. The backbone network uses large-scale pseudo-images to regress and select key points with high scores. Pixels around these key points are then divided into small scales, and the pseudo-images are regenerated. Finally, all pseudo-images are superimposed to generate features with excellent representation capabilities. Specifically, S in the figure represents the scale of pillars. The red square represents key points, and its surrounding orange squares represent the chosen pillars

at the cost of increased training difficulty and inference time. In contrast, low-resolution voxel divisions exhibit shorter training time and higher efficiency but lack local details, which leads to lower detection accuracy. Density distinctions in specific areas may be noticeable in the 3D space of one point cloud frame, since most point cloud voxels exist at a fixed scale. Moreover, it is common to use farthest point sampling in voxel-based networks. This approach unifies the number of points in various pillars to one value, while pillars with fewer points are filled with zeros. However, this sampling method causes an excessive deficiency of abundant information. Thus, a single fixed scale is insufficient to express features with all required information, especially for distant objects with sparse point clouds or small objects such as pedestrians. To resolve this issue, we propose a two-stage method to fuse multi-scale information around key points for objects that are regressed in the first stage and used to select point clouds for multi-scale fusion in the second stage. The framework is shown in Fig. 4.8.

A center-based method is then employed to produce center points in the pseudo-images, with chosen pillars corresponding to these center points. Pillars are processed by the IMMF module, allowing information from the same scales and different modalities to interact. Afterward, pseudo-images are generated from these concatenated features and stacked to enhance feature expression.

Specifically, the original cloud is divided into two scales (S and $S/2$). Initially, we encode pillars using the proportion of S and generate a pseudo-image $I \in R^{H \times W \times 64}$ with a size of $H \times W$. The coordinates (x, y) are then transferred to the corresponding ground truth heat map. Key points in the ground truth are calculated by:

$$\begin{cases} C_x = \dfrac{x - x_{min}}{(x_{max} - x_{min}) \times h_w} \\ C_y = \dfrac{y - y_{min}}{(y_{max} - y_{min}) \times h_l}, \end{cases} \tag{4.15}$$

where C_x and C_y are coordinates of a key point in the ground truth; x and y are center coordinates of a 3D bounding box; x_{min}, x_{max}, y_{min}, and y_{max} are individual minimum and maximum values; and h_w and h_l are the length and width of the heat map. All ground truth key points in the heat map are then split using a Gaussian kernel function.

Key point coordinates are applied over a large-scale space, as shown in Fig. 4.8. The expression of these key points can be represented by a heat map, with a ratio four times smaller than the input pseudo-image. In addition, the width and height of pillars with scales of $S/2$ are decreased by a factor of 2 compared with pillars of scale S, scaling all corresponding coordinates by a factor of 8. Pillars are then selected near adjacent key points, and a fixed area of square sides (exhibiting the most vehicles) is identified for use in generating pseudo-images. Images on a larger scale are then transformed to the proper dimensions by adaptive pooling. This allows pseudo-images on both large and small scales to be concatenated to generate high-dimensional features. Finally, the backbone network is applied to the high-dimensional features to regress detected objects.

4.3.4 *Experiments*

This section presents the results of several experiments applying the proposed M^2-Fusion framework to the Astyx HiRes 2019 dataset, verifying its advantages for 3D object detection. The proposed method is evaluated using comparisons with seven mainstream detection methods, including PointRCNN, SECOND, PV-RCNN, PointPillars, Part-A^2, Voxel R-CNN,and multimodal fusion method MVX-Net [8]. Adequate ablation studies are also conducted to verify the effectiveness of our each proposed module, including data preprocessing (DP), IMMF, and CMSF. In addition, hyperparameter tuning is used to assess performance trends.

4.3.4.1 Dataset

Astyx HiRes 2019 [5] is an open-access database that includes 4D RaDAR data for use in 3D object detection. Its purpose is to provide high-resolution 4D RaDAR data to the research community. The set consists of 4D RaDAR frames, 16-line LiDAR data, and camera images with temporal and spatial calibration. The data were split using a ratio of 3:1 to ensure the training and test distributions were consistent. The 4D RaDAR and LiDAR point clouds typically included 1000 to 10,000 points and 10,000 to 25,000 points, respectively, with images exhibiting a resolution of 2048 × 618 pixels. The maximum range of point cloud data reached 200 meters for 4D RaDAR and 100 meters for 16-line LiDAR. The LiDAR point clouds were then transferred to the RaDAR coordinate system, since the 3D bounding boxes were labeled in the RaDAR point cloud. The training set mainly included cars and very few pedestrians and cyclists. Therefore, experimental evaluation was conducted only for the car category, as the number of objects outside this was too small. Official KITTI evaluation protocols were followed, in which an IoU threshold of 0.5 was used for cars. IoU threshold was the same for the bird's-eye view (BEV) data and the entire 3D evaluation set. These methods were compared using the mean average precision (mAP) as an evaluation metric.

4.3.4.2 Implementation Details

The range of x, y, and z was set to (0, 69.12 m), (−39.68 m, 39.68 m), and (−3 m, 1 m) following the point cloud configuration in the KITTI dataset, respectively. The fusion network consisted of a pillar extraction module (used to extract pillars from 4D RaDAR and LiDAR point clouds), a feature fusion module, and a backbone network. Two different scales were utilized in the CMSF module, with pillar volumes of [0.16, 0.16, 4] (m) and [0.08, 0.08, 4] (m) for S and $S/2$, respectively. The input channel size for the IMMF module was 10, and the output channel size was 64, while the backbone network consisted of three convolutional blocks and three deconvolutional blocks. The number of convolution layers, the step size, and the number of output channels were [3, 5, 5], [2, 2, 2], and [128, 256, 512] in the three convolutional blocks, respectively, and [3, 5, 5], [1, 2, 4], and [128, 128, 128] in the three deconvolutional blocks, respectively.

4.3.4.3 Training

The open-source OpenPCDetcode framework was utilized to construct the training set, and a single NVIDIA RTX 3090 was employed to train the model, using 160 epochs. Other training parameters were consistent with PointPillars.

4.3.4.4 3D Object Detection on Astyx HiRes 2019 Dataset

Existing fusion networks mainly focus on modalities with different data formats. However, 4D RaDAR and LiDAR have similar formats, so features from the two sensors can complement each other. This is the underlying principle for our proposed multimodal fusion network.

Table 4.8 Comparative results for mainstream algorithms applied to the Astyx HiRes 2019 dataset

			3D mAP(%)			BEV mAP(%)		
Modality	Methods	Reference	Easy	Moderate	Hard	Easy	Moderate	Hard
4D RaDAR	PointRCNN	CVPR 2019	14.79	11.40	11.32	26.71	18.74	18.60
	SECOND	SENSORS 2018	23.26	18.02	17.06	37.92	31.01	28.83
	PV-RCNN	CVPR 2020	27.61	22.08	20.51	49.17	39.88	36.50
	PointPillars	CVPR 2019	26.03	20.49	20.40	47.38	38.21	36.74
	Part-A^2	TPAMI 2021	14.96	13.76	13.17	26.46	21.47	20.98
	Voxel R-CNN	AAAI 2021	23.65	18.71	18.47	37.77	31.26	27.83
16-line LiDAR	PointRCNN	CVPR 2019	39.03	29.97	29.66	41.34	34.22	32.95
	SECOND	SENSORS 2018	51.75	43.54	40.72	55.16	45.63	43.57
	PV-RCNN	CVPR 2020	54.63	44.71	41.26	56.08	46.68	44.86
	PointPillars	CVPR 2019	54.37	44.21	41.81	58.64	47.67	45.26
	Part-A^2	TPAMI 2021	45.41	38.45	36.74	49.85	41.85	38.93
	Voxel R-CNN	AAAI 2021	52.26	44.08	40.06	53.94	44.54	40.43
Camera + 4D RaDAR	MVX-Net	ICRA 2019	13.20	11.69	11.43	23.57	20.36	19.04
Camera + 16-line LiDAR	MVX-Net	ICRA 2019	39.16	31.43	30.40	47.04	38.15	35.60
4D RaDAR + 16-line LiDAR	M^2-Fusion(ours)		**61.33**	**49.85**	**49.12**	**71.27**	**61.24**	**57.03**

Bold values highlight the significant improvement in performance

To establish baselines, seven mainstream algorithms were applied to the KITTI 3D object detection benchmark. We first compared the performance of PointRCNN, SECOND, PV-RCNN, PointPillars, Part-A^2, Voxel R-CNN,and multimodal fusion method MVX-Netfor 4D RaDAR, 16-line LiDAR, and camera data from Astyx. The results were shown in Table 4.8. The point-based methods including PointR-CNN and Part-A^2 exhibited lower performance than other voxel-based methods significantly. The main reason was that the point clouds of both 4D RaDAR and 16-line LiDAR were very sparse, which was not conducive to the extraction of point features. PointPillars achieved the best accuracy in both 4D RaDAR and 16-line LiDAR point cloud except PV-RCNN. However, PointPillars had significant advantages in network structure complexity and inference time than PV-RCNN. Thus, we chose the easily expanded PointPillars network as our baseline to verify the proposed method. These algorithms were trained and compared with our model to evaluate the effectiveness of the proposed method for 4D RaDAR and LiDAR fusion. These results demonstrated that our proposed M^2-Fusion method achieved the best results, with significant improvements over other methods. This included increases of 5.64% (3D mAP) and 13.57% (BEV mAP) over the baseline PointPillars model trained with 16-line LiDAR. Moreover, the inference time for M^2-Fusion is about 10 fps on a single RTX 3090 GPU.

4.3.4.5 Ablation Studies with M^2-Fusion

Ablation studies were also used to verify the effectiveness of each proposed model compared to the baseline (PointPillars), the results of which were shown in Table 4.9. The first column showed the experimental configuration, including the sensor modality, data preprocessing (DP), and two proposed modules IMMF and CMSF. Assessment difficulty in the second and third columns included three levels: easy, moderate, and hard. The following data comparisons were all evaluated in the moderate level. It was evident that the results for the baseline (PointPillars) using 4D RaDAR alone produced the lowest accuracy. After data preprocessing, the 3D mAP for RaDAR increased by 1.50%, and the BEV mAP increased by 3.62%. These results suggested that data preprocessing offered certain improvements in RaDAR detection accuracy, which demonstrated its effectiveness. The third line showed the results for 16-line LiDAR, and the fourth line showed the results for 4D RaDAR fused with LiDAR using the direct feature concatenation. The fusion method got the improvement of 0.12% in 3D mAP, while achieved the improvement of 8.08% in BEV mAP. This illustrated that 4D RaDAR could improve the BEV accuracy remarkably. The next line provided the fusion results adding the data preprocessing,

Table 4.9 The results of ablation applying M^2-Fusion to the Astyx HiRes 2019 dataset (the baseline is PointPillars)

Methods					3D mAP(%)			BEV mAP(%)		
4D RaDAR	16-line LiDAR	DP	IMMF	CMSF	Easy	Moderate	Hard	Easy	Moderate	Hard
✓					26.03	20.49	20.40	47.38	38.21	36.74
✓		✓			28.61	21.99	21.35	50.66	41.83	38.70
	✓				54.37	44.21	41.81	58.64	47.67	45.26
✓	✓				54.25	44.33	43.24	66.05	55.75	54.67
✓	✓	✓			54.55	45.16	44.40	66.05	57.18	55.59
✓	✓	✓	✓		57.15	48.24	47.01	69.64	58.12	56.43
✓	✓	✓		✓	56.61	47.67	46.57	67.01	57.35	56.24
✓	✓	✓	✓	✓	**61.33**	**49.85**	**49.12**	**71.27**	**61.24**	**57.03**

Bold values highlight the significant improvement in performance

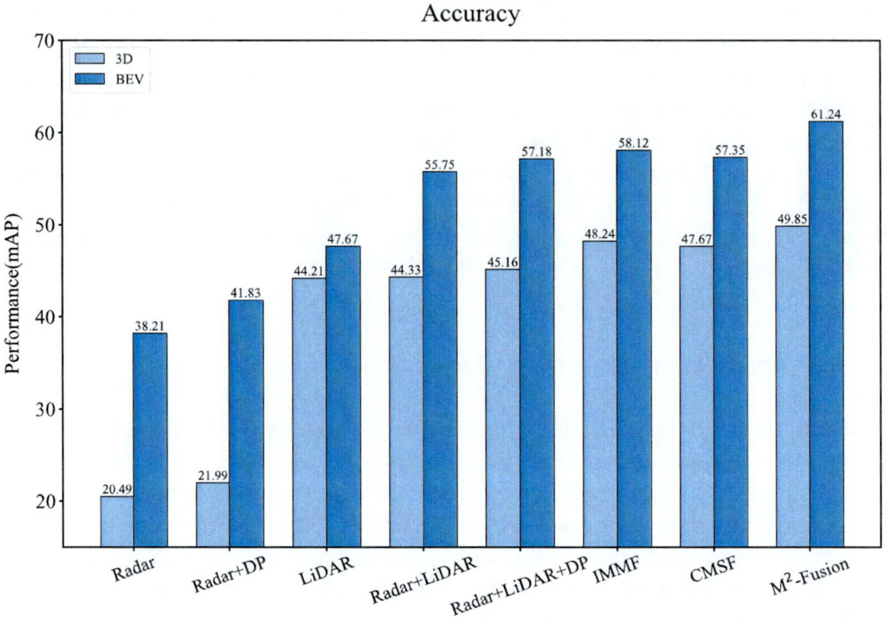

Fig. 4.9 Ablation results for M^2-Fusion applied to the Astyx HiRes 2019 dataset, with PointPillars serving as the baseline. The combination of CMSF and IMMF outperformed comparable models

illustrating the beneficial effectiveness. The sixth line showed the results for an added IMMF module, which produced increases of 3.08% (3D mAP) and 0.94% (BEV mAP) above the baseline (the fifth line, 4D RaDAR+16-line LiDAR+DP). The seventh line showed experimental results for an added CMSF module. The corresponding 3D and BEV mAP values increased by 2.51% and 0.17% above the baseline (the fifth line, 4D RaDAR+16-line LiDAR+DP), respectively. These above comparisons proved the remarkable effects of each proposed DP, IMMF, and CMSF module. Finally, we combined these three modules to produce M^2-Fusion, which achieved the best results with improvements of 5.64% in 3D mAP and 13.57% in BEV mAP compared to the traditional single LiDAR-based method (the third line, 16-line LiDAR). And the fusion of multiple modules acquired a more obvious improvement than a single module. These results were evident in Fig. 4.9 and confirmed the overall effectiveness of our proposed M^2-Fusion algorithm for 3D object detection.

We have the following observations through the above ablation studies: (1) The fusion of 4D RaDAR and 16-line LiDAR can achieve better results than the single-mode method. (2) The data processing of 4D RaDAR can correct data and suppress the influence of noise. (3) The attention-based interaction between 4D RaDAR and 16-line LiDAR can take advantage of each modality and enhance the perception ability. (4) Multi-scale feature extraction can obtain richer feature information and improve detection accuracy. Finally, our proposed M^2-Fusion method aggregates the above advantages to improve the detection accuracy significantly.

4.3.4.6 Accuracy Comparison Experiments at Different Ranges

4D RaDAR usually has a more extended range capability than LiDAR. To verify the object detection performance, we conducted the accuracy comparison experiments at different ranges for 4D RaDAR, 16-line LiDAR, direct fusion, and our proposed M^2-Fusion, respectively. The results were shown in Table 4.10 and the intuitive line chart in Fig. 4.10. The same method PointPillars was used following the above experiments in Table 4.9.

In terms of overall accuracy, the detection accuracy of 4D RaDAR was lower than that of LiDAR, but as the range increased, the detection accuracy gap between the two became smaller and smaller. In addition, after the detection range exceeded 30 m, the BEV detection accuracy of 4D RaDAR surpassed that of LiDAR, and the accuracy after 50 m increased by 10.45%, showing a significant effect. It showed that LiDAR had high accuracy in short-range detection, while 4D RaDAR had advantages in long-distance detection. After the direct fusion of the two sensors according to the feature concatenation method in Table 4.9 (line 4), the overall and most ranges' accuracy was increased, but the accuracy of the close range within 50 m was not significantly improved than that of LiDAR. In contrast, the detection accuracy above 50 m was improved by 3.41% and 15.13% on 3D and BEV mAP, respectively. It illustrated that fusion could improve the accuracy of remote detection. Our proposed M^2-Fusion method could significantly improve the detection accuracy within each range, and the overall accuracy of 3D and BEV was improved by 5.52% and 5.49%, respectively, which proved the effectiveness of the proposed method.

4.3.4.7 Parameter Comparison Experiment

High-resolution pseudo-images were reshaped using a unified scale to fuse point clouds of different resolutions. A variety of methods were available for reconstructing pseudo-images that could maintain the integrity of information to the extent possible prior to reconstruction. Various methods utilizing a standard size (when transforming pseudo-images from different scales to the same scale) were compared in Table 4.11. These results demonstrated the effectiveness of adaptive pooling, which outperformed CNN and max pooling methods and proved to be highly compatible with our network by outputting features of multiple dimensions.

Since the network in the CMSF module predicts key points with probabilistic scores, key points with low scores could lead to false detection. As such, we set a score threshold to remove these points. The influence of this threshold was tested using different values, the results of which were shown in Table 4.12. As shown, the

Table 4.10 Comparative results of 3D object detection using single modality and multi-fusion methods applied to different ranges. "Inf" means infinity

Modality	Methods	3D mAP(%)				BEV mAP(%)			
		Overall	0–30 m	30–50 m	50 m–Inf	Overall	0–30 m	30–50 m	50 m–Inf
4D RaDAR	PointPillars	20.49	34.06	14.76	6.98	38.21	52.08	28.81	19.54
16-line LiDAR	PointPillars	44.21	71.90	21.25	9.09	47.67	76.07	24.86	9.09
4D RaDAR + 16-line LiDAR	Direct fusion	44.33	67.67	21.50	12.50	55.75	77.92	35.64	24.22
4D RaDAR + 16-line LiDAR	M^2-Fusion(ours)	**49.85**	**77.26**	**27.36**	**15.56**	**61.24**	**83.73**	**42.08**	**27.68**
4D RaDAR + 16-line LiDAR	Improvement	+5.52	+9.59	+5.86	+3.06	+5.49	+5.81	+6.44	+3.46

Bold values highlight the significant improvement in performance

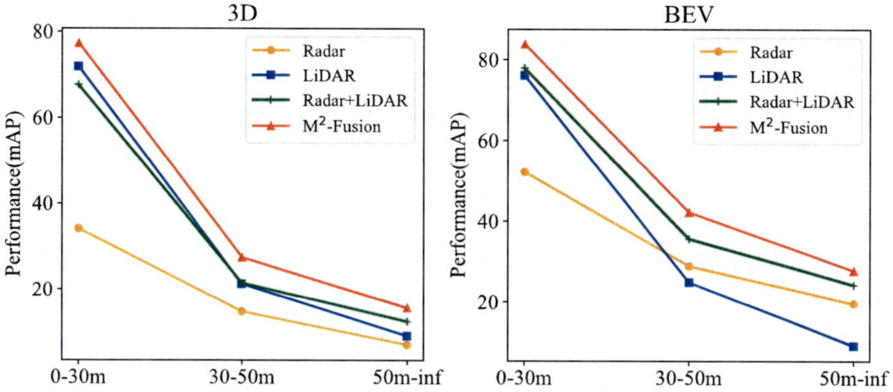

Fig. 4.10 Comparative results of 3D object detection in different ranges. Our proposed M^2-Fusion achieves the best accuracy in all ranges, especially remote detection

Table 4.11 A Comparison of different methods used to transform pseudo-images from different scales. MP and AMP mean the max pooling and adaptive max pooling, respectively

	3D mAP(%)			BEV mAP(%)		
Methods	Easy	Moderate	Hard	Easy	Moderate	Hard
CNN	58.01	47.93	46.57	69.19	57.10	55.58
MP	59.67	48.74	46.89	70.34	58.02	56.60
AMP	**61.33**	**49.85**	**49.12**	**71.27**	**61.24**	**57.03**

Bold values highlight the significant improvement in performance

Table 4.12 A Comparison of different scores for key point selection methods

	3D mAP(%)			BEV mAP(%)		
Score	Easy	Moderate	Hard	Easy	Moderate	Hard
0.2	57.65	47.71	45.10	70.05	56.80	55.61
0.4	56.72	46.90	45.65	70.56	57.68	57.02
0.6	**61.33**	**49.85**	**49.12**	**71.27**	**61.24**	**57.03**
0.8	57.07	48.20	46.75	67.59	57.79	56.39

Bold values highlight the significant improvement in performance

model achieved the best results for a score threshold of 0.6, suggesting that smaller thresholds introduced errors, while larger thresholds filtered out too much data.

Table 4.13 provided the results of comparison experiments involving concatenation at different scales in the CMSF module. Relying on prior experience, we selected two pillar scales, [0.16, 0.16, 4], and [0.08, 0.08, 4], with a baseline pillar scale of [0.16, 0.16, 4]. The use of different pillar scales had a direct impact on the experimental results, as the accuracy was higher for smaller scales within a specific range. As seen in Table 4.13, multi-scale fusion of [0.16, 0.16, 4] and [0.08, 0.08, 4] achieved the better results, producing a 1.61% increase in 3D mAP and a 3.12% increase in BEV mAP over the baseline in the moderate level. This experiment

Table 4.13 The results of scale parameter verification experiments using the CMSF module in our proposed M^2-Fusion method

	3D mAP(%)			BEV mAP(%)		
Scale(m)	Easy	Mod.	Hard	Easy	Mod.	Hard
0.16	57.15	48.24	47.00	69.64	58.12	56.43
0.16+0.08	**61.33**	**49.85**	**49.12**	**71.27**	**61.24**	**57.03**

Bold values highlight the significant improvement in performance

demonstrated that feature representations could be enhanced by fusing multiple scales.

4.3.4.8 Visualization Experiments

To observe the effects of modal fusion more intuitively, we provided tree visualization experiments including precision-recall (PR) curve comparison, feature map visualization, and qualitative detection results. Figure 4.11 showed the precision-recall (PR) curves of our proposed M^2-Fusion and the direct fusion of 4D RaDAR and LiDAR based on PointPillars. It revealed that our method significantly improved accuracy in easy, moderate, and hard levels. The visualization of feature maps for 4D RaDAR and LiDAR data using the baseline (PointPillars) and our proposed M^2-Fusion was shown in Fig. 4.12. It was evident that feature maps using our method contained more information. These results illustrated that multimodal interactive fusion enhanced network feature extraction capabilities and made extracting long-

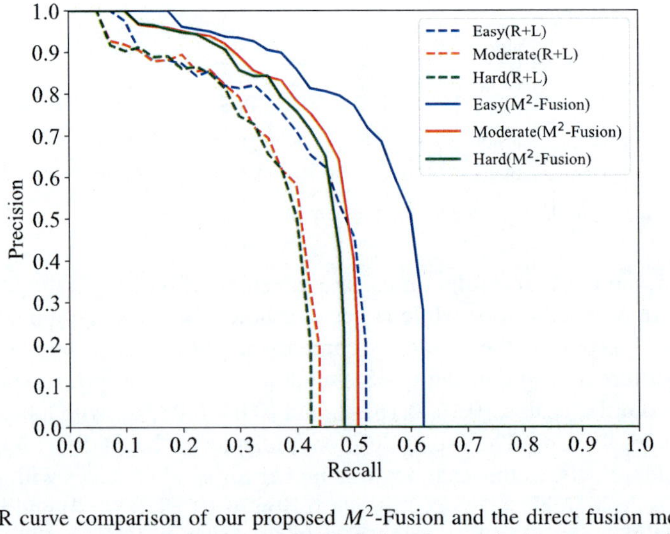

Fig. 4.11 PR curve comparison of our proposed M^2-Fusion and the direct fusion method based on PointPillars. "R+L" means RaDAR and LiDAR fusion. Our method surpasses the direct fusion method in easy, moderate, and hard levels observably

(a) (b) (c) (d)

Fig. 4.12 Feature maps for (**a**) 4D RaDAR data processed using PointPillars, (**b**) 4D RaDAR processed by M^2-Fusion, (**c**) 16-line LiDAR processed by PointPillars, and (**d**) 16-line LiDAR processed by M^2-Fusion

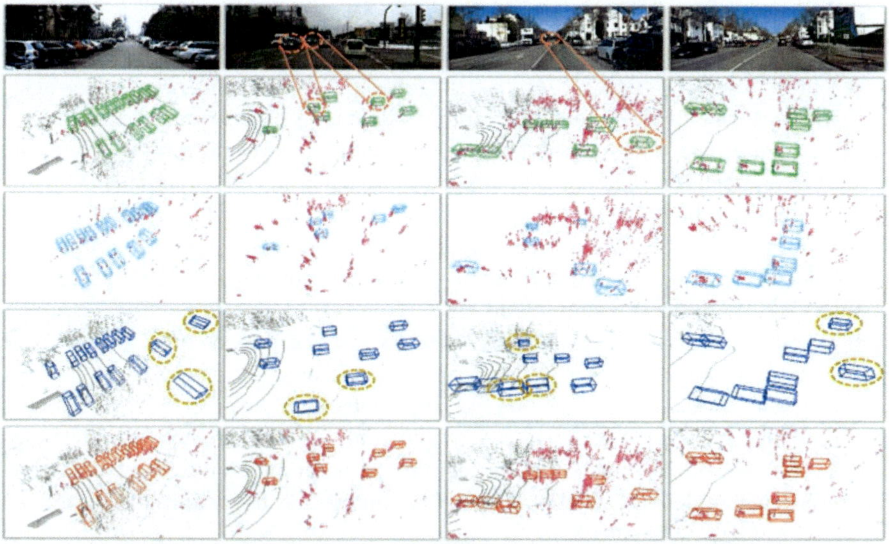

Fig. 4.13 Qualitative results from the Astyx HiRes 2019 dataset. The LiDAR point cloud is gray, and the 4D RaDAR point cloud is pink. The first row shows RGB images; the second row provides the ground truth; and the green bounding boxes surrounded by dotted orange circles indicate missed detections. The third row shows detection results for 4D RaDAR; the fourth row shows results for 16-line LiDAR; and the blue bounding box surrounded by dotted brown circles denotes false detections. The last row shows results produced by the proposed M^2-Fusion method

range features more accessible. We also provided a qualitative visualization of the final detection results for single RaDAR, LiDAR, and M^2-Fusion in Fig. 4.13, where green boxes denoted the ground truth. We observed that the baseline (PointPillars) method using only 4D RaDAR produced many false detections, as several trees and signal lights were considered cars. The baseline results using only LiDAR were somewhat better. However, objects located at large distances have not been detected effectively due to the low density of LiDAR points in these regions. In contrast, our proposed M^2-Fusion method achieved the best results and reduced false detections significantly.

4.4 Challenges and Prospect

The advantage of multimodal fusion is that data naturally exists in multiple modes. Therefore, if we can carry out multimodal fusion, we can obtain more comprehensive information. In the past time, 3D object detection based on multimodal fusion has achieved great success. There are many excellent improvement algorithms based on LiDAR and image fusion, LiDAR and RaDAR fusion, LiDAR and image fusion, and so on. Although 3D object detection based on multimodal fusion has developed rapidly and achieved remarkable detection results, many unsolved problems still exist.

- Difficulty of data fusion:

Different from single-mode data, multi-mode data is more comprehensive data, such as image data and RaDAR point cloud data in target detection, image and voice in the video, etc. The image data contains rich color information and texture information, but the deficiency is the lack of depth information. That is, the image information can only represent two-dimensional information. The representation of point cloud data is disordered and sparse, so the two-dimensional CNN perceptron cannot directly process point cloud data. However, point cloud data contains rich geometric structure and depth information in three-dimensional space, so the information of image and point cloud data is complementary in theory.

At present, in two-dimensional image target detection, the deep learning network detection model is designed based on CNN. In contrast, there are networks designed by MLP, CNN, GCN, and other infrastructure in point cloud target detection. Therefore, which network to choose in the fusion process needs further research.

- Difficulty of feature extraction:

Multimodal feature extraction proposes the redundancy between modes by using the complementary information of multimodal to learn better feature representation. Since multimodal feature fusion is done, the corresponding fusion can only be carried out by the connection between image information and point cloud information. In the feature layer or input layer, the relationship between the image and the point cloud comes from cognition, that is, no matter what kind of single-mode sensor, scanning the same object at the same time is instantaneous representation information of the object, in which the representation forms of different sensors are different. Still, the commonness of multi-mode data information is the instantaneous absolute coordinates of the object. Therefore, the connection and link between the image and the point cloud are the instantaneous absolute coordinates of the object.

The different world coordinate systems of the camera and the RaDAR result in different instantaneous relative coordinates of the data information. However, due to the commonness of the instantaneous absolute coordinates of the object, the coordinate transformation between the two sensor coordinate systems can be easily obtained only through the position transformation matrix of the RaDAR and the camera. In this way, the scanned object can be used as a link by extracting the coordinate features under the two sensors.

However, in the process of feature extraction, the size of the feature map or domain may change so that the most original coordinates will change to some extent, resulting in partial information loss or redundancy, which is also an urgent problem to be studied. Similarly, in the process of information complementarity, how to eliminate redundant information also needs further research.

- Difficulty of data registration:

From the perspective of coupled structure, multimodal fusion means that the two modes are related at some levels. The fusion behavior requires the comprehensive utilization of redundant and complementary features. Therefore, in the process of real data processing, there will be a problem with data registration. For example, for the problem of sensor angle of view, the information obtained by the camera is the information obtained from a cone of view of small hole imaging. At the same time, the LiDAR is the information obtained in the real 3D world, which makes the information data representation of the same target object very different, so it is necessary to consider the problem of data registration and alignment under different modal resolutions. This difficulty is the problem that multimodal fusion will encounter. The image information is dense and regular, but the information of the point cloud is sparse and disordered. Therefore, using one rich modal information to assist another relatively poor modal for data registration and collaborative learning needs further research.

In recent years, with the rapid development of artificial intelligence, multimodal fusion has gradually become a research hotspot in 3D target detection. The current research on modal fusion technology has promoted the emergence of many new multimodal algorithms and expanded the application scope of multimodal fusion learning. These models and algorithms have advantages and disadvantages and can play their advantages and roles in different fields. At present, the most widely used field of three-dimensional target detection based on multimodal fusion is automatic driving. In the process of automated driving, many sensors are needed for collaborative optimization. It is difficult for a single mode to carry out reasonable visual perception. It can be introduced into infrared, visible RaDAR, and LiDAR for joint optimization. As an automatic intelligent driving vehicle technology integrating the national industrial manufacturing field and new technologies, multimodal fusion deep learning is expected to make great progress in the future. In the next step, we can further study the difficulties of data fusion, the selection of feature extraction, data registration, multimodal combination evaluation criteria, modal generalization ability, and other insufficient research problems, deeply explore the difficult problems such as cross-modal transfer learning and nonconvex optimization, and promote the application of this technology in some new fields of 3D target detection.

Finally, the author believes that although multimodal fusion 3D target detection has made significant progress and has become an essential branch of the development of target detection, the ultimate goal is to build an agent that can perceive multimodal information and use the relationship between different modes to improve its cognitive detection ability. At present, the research on multimodal fusion 3D target detection is still in its infancy, which faces both significant challenges and great opportunities.

4.5 Summary

Self-driving cars are becoming more popular nowadays, which transport with their intelligence and take appropriate actions at the adequate time. Safety is a critical factor in the driving environment. A simple failure of action can cause many fatalities. Computer vision plays a significant part in achieving this. It helps the autonomous vehicle to perceive the surroundings. Detection is a prevalent technique in helping to capture the surrounding of an autonomous car. At the same time, tracking also has an essential role in this by providing dynamic of detected objects. Autonomous cars combine various sensors such as RaDAR, LiDAR, sonar, GPS, odometry, and inertial measurement units to perceive their surroundings. Driver-assistive technologies like adaptive cruise control, forward collision warning system, and collision mitigation by breaking ensure safety while driving. Perceiving the information from the environment include setting up sensors on the car. These sensors will collect the data it sees, and this will be further processed for taking action. The sensor system can be a single sensor or multiple sensors. Different sensors have different strengths and weaknesses, which makes the combination of them important for technologies like autonomous driving. Each sensor will have a limit of accuracy on its readings, so a multi-sensor system can help to overcome these defects. This thesis is an attempt to develop a multi-sensor multi-object tracking method to perceive the surrounding of the ego vehicle. We present several fusion methods to fuse data from multiple sensors, promoting detection accuracy in a particular range. In future work, multi-sensor fusion will be the mainstream in object detection.

References

1. Chen, X., Ma, H., Wan, J., Li, B., Xia, T.: Multi-view 3D object detection network for autonomous driving. In: 2017 IEEE Conference on Computer Vision and Pattern Recognition (CVPR), pp. 6526–6534 (2017). https://doi.org/10.1109/CVPR.2017.691
2. Ku, J., Mozifian, M., Lee, J., Harakeh, A., Waslander, S.L.: Joint 3D proposal generation and object detection from view aggregation. In: 2018 IEEE/RSJ International Conference on

Intelligent Robots and Systems (IROS), pp. 1–8 (2018). https://doi.org/10.1109/IROS.2018.8594049
3. Kuang, H., Wang, B., An, J., Zhang, M., Zhang, Z.: Voxel-FPN: Multi-scale voxel feature aggregation for 3D object detection from LiDAR point clouds. Sensors **20**(3), 704 (2020)
4. Lang, A.H., Vora, S., Caesar, H., Zhou, L., Yang, J., Beijbom, O.: PointPillars: fast encoders for object detection from point clouds. In: 2019 IEEE/CVF Conference on Computer Vision and Pattern Recognition (CVPR), pp. 12,689–12,697 (2019). https://doi.org/10.1109/CVPR.2019.01298
5. Meyer, M., Kuschk, G.: Automotive Radar Dataset for Deep Learning Based 3D Object Detection. In: 2019 16th European RaDAR Conference (EuRAD), pp. 129–132 (2019)
6. Shi, S., Guo, C., Jiang, L., Wang, Z., Shi, J., Wang, X., Li, H.: PV-RCNN: Point-voxel feature set abstraction for 3D object detection. In: 2020 IEEE/CVF Conference on Computer Vision and Pattern Recognition (CVPR), pp. 10,526–10,535 (2020). https://doi.org/10.1109/CVPR42600.2020.01054
7. Shi, S., Wang, X., Li, H.: PointRCNN: 3D object proposal generation and detection from point cloud. In: 2019 IEEE/CVF Conference on Computer Vision and Pattern Recognition (CVPR), pp. 770–779 (2019). https://doi.org/10.1109/CVPR.2019.00086
8. Sindagi, V.A., Zhou, Y., Tuzel, O.: MVX-Net: Multimodal VoxelNet for 3D object detection. In: 2019 International Conference on Robotics and Automation (ICRA), pp. 7276–7282 (2019). https://doi.org/10.1109/ICRA.2019.8794195
9. Vora, S., Lang, A.H., Helou, B., Beijbom, O.: PointPainting: sequential fusion for 3D object detection. In: 2020 IEEE/CVF Conference on Computer Vision and Pattern Recognition (CVPR), pp. 4603–4611 (2020). https://doi.org/10.1109/CVPR42600.2020.00466
10. Yan, Y., Mao, Y., Li, B.: SECOND: sparsely embedded convolutional detection. Sensors **18**, 3337 (2018). https://doi.org/10.3390/s18103337

Chapter 5
Multi-Sensor Scene Segmentation

Abstract Semantic scene understanding is crucial for autonomous driving, robot navigation, and intelligent scene perception tasks. Semantic segmentation can identify and locate targets at the pixel level, providing essential situation awareness for navigation tasks. This chapter focuses on the most commonly used semantic segmentation task in autonomous driving, namely, lane line detection. From the perspective of multimodal fusion, the task of semantic segmentation is revisited. Due to the gap in different modalities, the alignment among of multi-sensor data is challenging for fusion, especially for scene segmentation. Several novel lane line detection algorithms are designed based on attention mechanisms, mutual information, and an end-to-end approach to solve this problem. This chapter first discusses the different fusion architectures and then utilizes the attention mechanism to achieve better results. Further, an adaptive fusion strategy is proposed to address robustness and bias in fusion. After experimental verification, the proposed methods have achieved good performance in segmentation tasks.

5.1 Background

Object segmentation basically separates the objects of interest from the background, and structure labeling assigns a unique object class label to each pixel of the image. Therefore, the object segmentation and structure marking labeling have the same target, and while the focus of object segmentation is to detect the boundary of the object, the focus of structure marking is to detect the region of the object.

Some early semantic segmentation methods use small flat classifiers to classify the central pixels and then use conditional random fields (CRFs) for smooth prediction. In the random decision forest, they use structure from motion features as well as appearance-based features. In the past few years, with the success of deep learning in image classification, deep neural networks (DNN) for semantic segmentation have developed rapidly.

The rapid development of neural networks improves the performance of AI-driven image segmentation models, but the ability to accurately and steadily deal with challenging outdoor environments and conditions is still not enough and needs

X. Zhang et al., *Multi-sensor Fusion for Autonomous Driving*,
https://doi.org/10.1007/978-981-99-3280-1_5

to be further improved. Moreover, if only a separate sensor is used for scene understanding, it is easy to be limited by the sensor, so using a multi-sensor such as an RGB camera, depth camera, and LiDAR sensor can make the sensing model more robust under various conditions.

Generally speaking, to establish an accurate and reliable multi-sensor segmentation framework, we need to pay attention to the following two factors: The first is to explore different imaging methods. The second is to study the fusion method of complementary information in different modal images. There are several types of fusion strategies for learning multimodal imaging. The first is early fusion, which fuses data into the input of a machine learning algorithm; the second fusion type is late fusion, in which the features of each mode are learned separately and fused at a deeper level of neural network or other classifiers. Different imaging methods and fusion methods play a decisive role in the segmentation effect.

5.2 Introduction

Nowadays, the large majority of state-of-the-art algorithms for lane line segmentation use machine learning technology to some extent. However, Tang et al. [19] pointed out it is still a challenge to develop a robust detector under unlimited conditions. Visual lane line detection is easily affected by weather and light intensity, and excessive post-processing also limits the detection efficiency of the model.

The current approaches for lane line segmentation can be divided into three types: camera-based models, LiDAR-based models, and fusion models. Both camera-based and LiDAR-based methods have distinct advantages and disadvantages. To optimally consider the advantages of different sensors in accurately describing the external environment, it is necessary to fuse the data from different sensors [22]. Multimodal fusion involves combining data from different sensors to perform multi-stage feature fusion at different levels [13]; through multimodal fusion, this method can improve the accuracy of lane line segmentation. Notably, the redundancy and fault tolerance of the system and the speed and accuracy of decision-making can be improved [2].

Though existing multimodal methods can perform well in most scenes, their fusion strategies may fail severely in some abnormal scenarios [9]. For instance, bad weather like rainy and foggy days can put obstacles in the way of the camera's work. However, in multimodal perception fusion, the questions of "when to fuse" and "how to fuse" remain unanswered [7]. The current fusion strategies only add feature maps for two sets of sensor data at a fixed ratio and then input the overall result into the network. However, using a fixed ratio is inflexible and unreasonable.

The use of multimodal sensors has become increasingly common to achieve robust lane line segmentation [17]. We have found empirically that the contributions of different extracted feature channels are different and the direct fusion method does not use the contribution of feature channels as a way to assign weights.

Therefore, we use the channel attention mechanism to obtain the cross-channel local interaction information, and the weights of feature channels are assigned to represent the contributions of different feature channels, as described in Sect. 5.3.

Due to the varied weather and road conditions, the real scenes can be far more complicated than those in the training dataset. That constructs a non-ignorable challenge for multimodal fusion models that obey fixed fusion modes, especially for autonomous driving. To address the problem, we leverage mutual information for adaptive modal selection in fusion, which measures the relationship between the input and target output, as described in Sect. 5.4.

A fast and robust lane detection algorithm is required for autonomous vehicles, as it can help the car's safety while driving. The fusion methods utilize the complementation of camera images and 3D point clouds to improve the perception of autonomous driving by the deep network.

Some research ignores the importance of modal balance. Dislocation integration and inadequate integration of different modalities will lead to uneven feature contribution and representation. Based on this, we proposed a real-time multimodality fusion model with an attention mechanism to fully exploit the deep feature complementary between different modalities and achieve better lane detection performance, as described in Sect. 5.5.

The relevant methods and experiments we proposed are introduced in the following sections. Section 5.3 introduces the channel attention module and related experiments that can be easily added to the fusion segmentation network. In Sect. 5.4, the adaptive multimodal fusion network MIMF and related experiments are introduced. In Sect. 5.5, the multimodal fusion method and related experiments in lane detection are introduced.

5.3 Attention in Multimodal Fusion Segmentation

To assess the contributions of the different feature channels of sensors, we introduce a novel multimodal fusion method and demonstrate its practical utility using LiDAR-camera fusion networks. Specifically, a channel attention module that can be easily added to a fusion segmentation network is proposed. In this module, we use the channel attention mechanism to obtain the cross-channel local interaction information, and the weights of feature channels are assigned to represent the contributions of different feature channels. To verify the effectiveness of the proposed method, we conducted experiments on two types of feature fusion with the KITTI benchmark and the A2D2 dataset. As shown in Fig. 5.1, the results verify the benefits of the proposed fusion method. Our model achieves precise edge segmentation with a 5.59% gain in precision and a 2.12% gain in F2 score compared to the values of the original fusion method. We believe that we have introduced a new optimization idea for multimodal fusion.

Fig. 5.1 Results on three state-of-the-art algorithms (LaneNet, SCNN, ENet-SAD) and our method

5.3.1 Network Architectures

In this part, we introduce the structure of multimodal fusion at different stages and combine the CFECA module with the baseline to complete the lane line segmentation.

In the task of lane segmentation, the number of positive samples is much less than the number of negative samples. This can be considered unbalanced data. We establish a baseline model based on U-Net. It is composed of four blocks in the encoder and five blocks in the decoder, two of which are ResNet-34 blocks. The last four layers use transposition convolution, and the rest are convolution blocks. Note that the ResNet-34 used here is the whole ResNet module, and the strategy of importing pre-training parameters is not adopted, but it is initialized. All convolution blocks have a batch normalization layer and a RELU layer behind the convolution layer, and all kernel sizes are 3×3. Each block in the encoder is linked to the same block in the decoder so that less spatial information is lost during the down-sampling process.

Point cloud data is different from image data. Whether it is a two-dimensional image or a three-dimensional image, the image data fills the entire area, and each pixel in the two-dimensional image has a corresponding value, such as gray value, RGB value, etc. A three-dimensional image discretizes three-dimensional space into voxels [9] and assigns points to voxels. The point cloud data is different from the image data due to the characteristics of the data generation process of scanning, and the sparse and irregular characteristics of the LiDAR point cloud will lead to empty voxels. Research have shown that [3] first complete the sparse point cloud and then use the height information as an additional filtering condition to superimpose the height and distance values on the reflectance to form a pseudo-three-channel image, which can effectively reduce the interference of noise [1]. To reduce the impact of non-uniformity and data loss caused by empty voxels, we apply k-NN ($k = 2$) search point cloud interpolation to search three closest points for each blank pixel

[14] and calculate the weighted average value according to the normalized pixel distance. At the same time, to reduce the impact of point cloud noise and use the height information, the height and distance values are superimposed on the reflection coefficient to form a pseudo-three-channel image. All the information in the pseudo-three-channel image comes from the point cloud. In the following sections, we will call such pseudo-three-channel images a "point cloud."

Deep neural networks represent features hierarchically. In the process of data processing, "when to fuse" is a question worthy of exploring. To this end, the character and function of early fusion, middle fusion, and late fusion need to be discussed.

(1) Early fusion: Early fusion can be called "pixel-level fusion," which combines raw or preprocessed sensor data and has low calculation requirements and low storage budget, but it is sensitive to spatiotemporal data, which is caused by imbalance between sensors. (2) Middle fusion: Middle fusion is a "feature-level fusion." One of the advantages of the middle fusion method is that it can flexibly choose the location of the fusion, but it is not easy to find the best method for the fusion of the middle layer under a specific network architecture. We explored the middle fusion of the camera image branch and the point cloud branch behind the second ResNet-34 block named Mid-1 and the block behind Mid-1 named Mid-2. (3) Late fusion: Late fusion is application-oriented fusion, which can be regarded as "decision-level fusion." The late fusion has high flexibility and modularity, but it has disadvantage on high computing cost and storage requirements. The errors generated by the two classifiers will affect the segmentation results.

The KITTI Road dataset is an unbalanced dataset. The accumulation of classification errors will cause some classification errors to be magnified after using the attention module, which will have a great impact on the segmentation results. Therefore, we do not use the fusion method of the attention mechanism for late fusion.

The previous fusion network structure achieves the basic LiDAR-camera fusion. However, in the lane line segmentation task, the proportion of lane lines is far less than that of background. In our network, a small number of lane line labels are positive samples, and a large number of background samples are negative samples. In contrast, it is more important to efficiently extract the features of the lane line.

The channel attention mechanism argues that the importance of different feature channels in the network is different. It gives each channel in the network a weight which represents the importance of this channel. However, the original fusion network did not consider that the importance of feature channels was different. In this work, we use the channel attention mechanism to assign weights to the feature channels of point clouds and camera images. Since each feature channel has different weights, it will have varying degrees of influence on the feature map output by the fusion layer. In general, the relatively important feature channels and their related channels will get a larger weight in the process of adaptive weighting, while the channels that make a negative contribution to the segmentation will be suppressed. After the weighting of the channel attention mechanism, the network layer behind the fusion layer will get a new feature map, which will be more effective for extracting the lane line features.

However, most existing methods are devoted to the development of more complex attention modules to obtain better performance, which inevitably increases the computational burden. Channel attention has great potential for improving the performance of deep convolutional neural networks. As mentioned in related work, ECA-Net [20] is not only excellent in performance but also more suitable for multimodal fusion architectures. Considering the high real-time requirements of lane line segmentation tasks, here we discuss the role of the ECA module in LiDAR-camera fusion and design the CFECA module to improve the fusion method; by modeling the importance of each feature channel, the feature channel with a large contribution to the lane line segmentation task is enhanced while suppressing the feature channel with a negative contribution.

To avoid loss of dimensionality while capturing local cross-channel interactive information, we chose to integrate the ECA module into the task of lane line segmentation and obtain a model with lower model complexity and smaller network parameters.

In the fusion network, let the output of a convolution block be $\mathscr{X} \in \mathbb{R}^{W \times H \times C}$, where W, H, and C are the width, height, and channel dimension. We performed a global average pooling on this feature map, i.e.,

$$g(\mathscr{X}) = \frac{1}{WH} \sum_{i=1,j=1}^{W,H} \mathscr{X}_{ij} \tag{5.1}$$

To find an effective method for cross-channel interaction utilization, only the interaction between each channel and its k-nearest neighbors is considered. Besides, all channels share the same learning parameters to further reduce model complexity and improve efficiency. Then the weight of y_i can be calculated as:

$$\omega_i = \sigma\left(\sum_{j=1}^{k} w^j y_i^j\right), y_i^j \in \Omega_i^k \tag{5.2}$$

where Ω_i^k represents the set of k adjacent channels of y_i.

The ECA module can be easily implemented by fast 1D convolution with the kernel size of k, i.e.,

$$\boldsymbol{\omega} = \sigma\left(\mathrm{C1D}_k(\mathbf{y})\right) \tag{5.3}$$

where C1D represents one-dimensional fast convolution.

For different stages of LiDAR-camera fusion, we add a channel attention module to the previous stitching operation, and the image branch and point cloud branch are weighted to output the fused feature map as the input of the next layer in the original

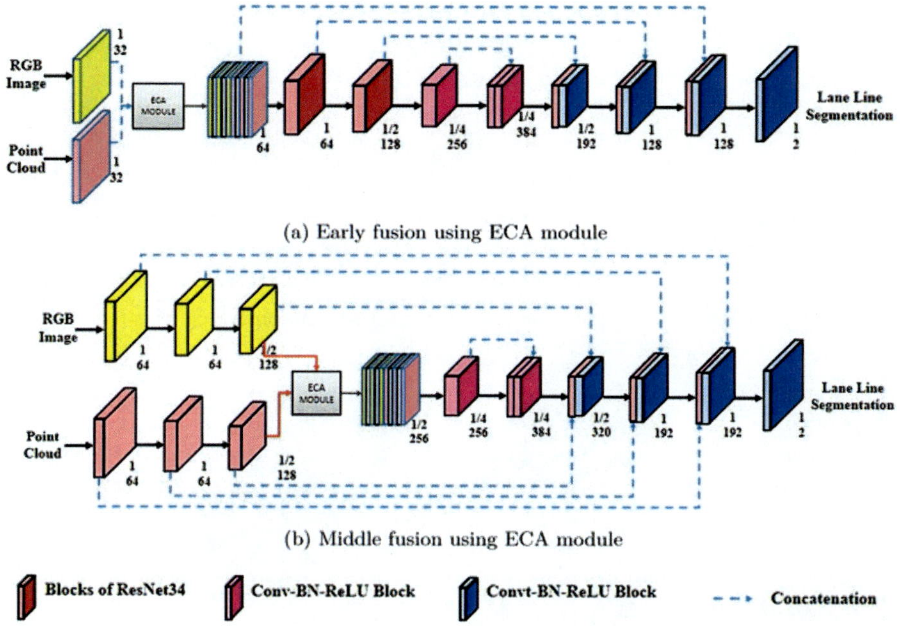

Fig. 5.2 Early fusion and middle fusion using ECA module. Mid-1 fusion and Mid-2 fusion are two stages of middle fusion, (**b**) shows the Mid-1 fusion, the Mid-2 fusion is the same as the fusion of Mid-1, and the position of Mid-2 is a convolution block behind Mid-1 fusion

network structure. The network structure after adding the ECA module is shown in Fig. 5.2.

Using the ECA module to process the result of the direct stitching of the camera image and the point cloud, to a certain extent, the local information of cross-channel is captured. However, this method can only merge the camera image branch and the LiDAR point cloud branch at the junction of the stitched channel. For other channels of the splicing result, the interaction between the guiding channel and the guided channel only comes from a single original branch. The key to multimodal fusion lies in whether the complementarity of multimodal information can be fully utilized. In the channel domain, cross-channel interaction is a fusion of information, and it also takes advantage of the complementarity of multimodal information. To make better use of the complementarity between the camera image and the point cloud data, and to use more fusion information to guide channel attention prediction, we have improved the previous attention module, as shown in Fig. 5.3.

We cross-splice the feature map of the camera image and the feature map of the point cloud and then perform the attention prediction of the channel. For the channel vector obtained by the global average pooling of cross-stitching, the attention prediction output of the ith channel is from the $i-(k-1)/2$ channel to the $i+(k-1)/2$ channel as guidance channels. The result is guided by the information interaction of the k channels. Without dimensionality reduction, by considering each

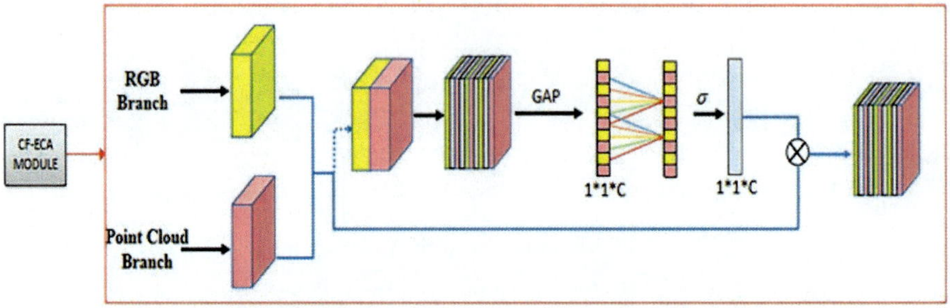

(a) Cross fusion efficient channel attention module

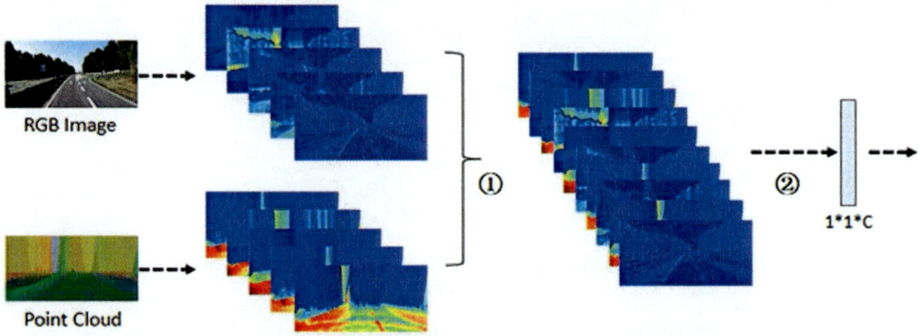

(b) Lidar and camera images cross stitch feature channels during the fusion process

Fig. 5.3 CFECA module and the fusion strategy; in (**b**), ① shows the cross-stitching of feature channels, and ② represents the weight calculation of feature channels after fusion

channel and its k neighbor channels, local cross-channel interaction information is captured. The k-value reflects how many neighbor channels participate in the prediction of a channel's attention. The interaction coverage between channels is related to the channel dimension. The k-value can be adaptively determined by the function related to the channel dimension.

$$C = \phi(k) = 2^{(\gamma * k - b)} \tag{5.4}$$

$$k = \psi(C) = \left| \frac{\log_2(C)}{\gamma} + \frac{b}{\gamma} \right|_{\text{odd}} \tag{5.5}$$

where $|t|_{\text{odd}}$ indicates the nearest odd number of t. In this chapter, we set γ and b to 2 and 1 throughout all the experiments. At the same time, k must be less than or equal to 9. The original k is an adaptive odd number.

The above method will reduce the information from the original modality by half. We empirically found that when we restore the number of guidance channels to $2k - 1$, which means the channel information from the original modal remains unchanged, only adding another channel information of another modal and cross-

(a) RGB Image (b) Lane line Lable (c) Point Cloud

Fig. 5.4 Some examples of KITTI (top row) and A2D2 (bottom row). The original point cloud is completed by k-NN ($k = 2$), which is a pseudo-three-channel image containing reflection intensity, depth, and height. We use the pseudo-three-channel image in the following experiments, and we still call such a pseudo-three-channel image a "point cloud" as shown in (**c**). The upper part of A2D2's point cloud is cut in half to reduce the interference of irrelevant factors

Table 5.1 Basic information of three lane line segmentation datasets

Name	Frame	Train	Validation	Test	Resolution	Road type
KITTI	383	228	40	115	1242×375	Urban, rural, and highway
KITTI-aug	3331	2736	480	115	1242×375	Urban, rural, and highway
A2D2	470	282	47	141	1920×1208	urban

splicing them in the original order will maximize the use of the correlation between adjacent channels and the information complementarity of different modals.

5.3.2 Experiments and Discussion

Figure 5.4 shows several frames of three datasets that we use in our experiments. They are KITTI, KITTI-aug, and A2D2. KITTI is currently one of the most important test sets in the field of autonomous driving. It includes synchronized camera images and LiDAR point clouds with calibration parameters and ground truth. To verify our proposed method and models, we selected 383 sets of data without intersections or forward lines in the KITTI road detection track. We use strategies such as cropping, brightness transformation, and noise increase on the KITTI dataset to obtain a dataset with 12 times the number of KITTI, called KITTI-aug, to verify the robustness of the proposed method. As to A2D2, it combines only one 8-line and two 16-line LiDARs, while KITTI uses a 64-line Velodyne to generate point clouds. We selected 470 pairs of them for testing. The details of the datasets are shown in Table 5.1. For all datasets, we use 60% of the data as the training set, 10% for verification and the rest for testing. The difference in LiDARs causes the gap in performance, and the amount of data is also an important factor limiting model performance. Therefore, the main comparison experiment will be carried out on KITTI-aug instead of KITTI or A2D2.

In our work with KITTI-aug, we first added pixel-level lane line annotations to the KITTI dataset to better solve the semantic segmentation problem. Compared with the ApolloScape dataset and the TuSimple dataset, we filter out confusing lanes, such as the markings on the sidewalk. To extract the lane line features more accurately, we carefully mark the lane lines and ignore any marks behind obstacles such as vehicles and poles to reduce noise interference and then perform data enhancement to get the KITTI-aug dataset. We use a two-branch structure to separately extract the features of the camera image and the LiDAR point cloud. After the multimodal data fusion process must meet the feature alignment requirements, the initial size of the camera images and the corresponding point cloud images are 1242×375. After cropping, reshape them to the size of (256, 512) in the same way, and input them into the network.

All networks are implemented on a standard PC with 10 GB of RAM, a three-core E5-2678v3 CPU, and a NVIDIA GTX 1080 Ti GPU with 11 GB of memory under Ubuntu 16.04 using PyTorch. We use Adam optimization algorithm to train the network in an end-to-end manner and use the *torch.nn.NLLLoss2d* in PyTorch; the cyclical decayed learning rate lr is shown below:

$$lr = 2^{\lfloor epoch/50 \rfloor} \times 0.8^{\lfloor epoch/10 \rfloor} \times lr_0, \quad lr_0 = 0.0001 \tag{5.6}$$

where $lr_0 = 0.0001$ is the initial learning rate, the training epoch and batch size are set to 200 and 4, and every 5 epochs will be evaluated on the validation set. If there is any improvement, the network weights will be saved.

Recall and precision are both important indicators for measuring the performance of the model. Recall reflects how many lane line pixels we have correctly predicted, and precision represents how many real positive samples are in our prediction results. An excellent lane line segmentation model must not only find the lane line but also reduce the possibility of misjudgment as much as possible to ensure the safety of the vehicle. Besides, there are evaluation measures used in the following comparisons. They are F1-measure, F2-measure, the accuracy of the overall prediction named Acc, and the processing speed index FPS for lane line detection.

We also report the FP and FN of the different algorithms in Table 5.4 so that we can evaluate the performance of the model from more aspects. It can be seen from Table 5.2 that using the dataset after data enhancement of KITTI, the fusion model at different stages has achieved a certain degree of performance improvement. It is observed that early fusion's precision and Mid-1 fusion and Mid-2 fusion's recall have obtained a stable improvement. However, A2D2 combines only one 8-line and two 16-line LiDARs, while KITTI uses a 64-line Velodyne to generate point clouds. The experimental results show that compared with the LiDAR point cloud that is sparse, the dense and rich LiDAR point cloud contains more depth information, which has obvious advantages when testing the multimodal fusion model. It is observed that the overall performance of early fusion is relatively stable. Affected

Table 5.2 Comparison of different fusion stages on three datasets. "REC" denotes "Recall" and "PRE" denotes "Precision," and we use the same abbreviation in the following sections

Dataset	Fusion stage	REC	PRE	F1	F2	Acc	FPS
KITTI	Early	93.34	42.89	57.89	74.22	98.16	83.1
	Mid-1	86.97	51.94	61.72	73.44	98.56	61.9
	Mid-2	86.06	51.73	61.07	72.74	98.42	54.3
KITTI-aug	Early	92.78	50.28	64.46	78.30	98.59	80.4
	Mid-1	92.19	51.56	65.27	78.49	98.65	61.5
	Mid-2	91.82	50.20	63.88	77.72	98.34	54.9
A2D2	Early	87.06	32.09	45.06	61.57	98.87	72.3
	Mid-1	89.51	24.06	36.20	53.91	98.22	58.3
	Mid-2	89.47	24.02	36.19	53.66	98.18	57.6

Table 5.3 Comparison of three stages of fusion with and without ECA/CFECA

Fusion stage	Category	REC	PRE	F1	F2	Acc	FPS
Early	–	92.78	50.28	64.46	78.30	98.59	80.4
	ECA	93.42	48.45	63.00	77.63	98.49	78.3
	CFECA	92.40	54.37	67.70	80.05	98.80	61.1
Mid-1	–	92.19	51.56	65.27	78.49	98.65	61.5
	ECA	91.08	50.80	64.28	78.09	98.47	64.0
	CFECA	92.41	53.11	66.71	79.59	98.73	59.7
Mid-2	–	91.82	50.20	63.88	77.72	98.34	54.9
	ECA	91.68	49.92	63.75	77.34	98.28	54.6
	CFECA	92.16	52.21	66.08	79.35	98.71	53.7

by the quality of the LiDAR point cloud, the accuracy of Mid-1 fusion and Mid-2 fusion on A2D2 drops sharply, and the performance of Mid-1 fusion is slightly higher than that of Mid-2 fusion. In terms of the inference time required by the algorithm, it can be clearly seen that the inference time of early fusion is the smallest and the inference time required by Mid-1 fusion is slightly smaller than that of Mid-2 fusion. This suggests that as the fusion stage moves later, the camera image branch and the LiDAR point cloud branch each extract deeper features and then merge them in middle fusion that takes more time than those operations in the early fusion.

The results of our method at different fusion stages are shown in Table 5.3. Among them, the accuracy of CFECA in early fusion reaches 54.37%, which is much higher than other models. The F2 score reaches 80.05%, which is the only model with an F2 score over 80%. It can be seen that the overall performance is also the best in it. We used the ECA module directly and the improved CFECA module for the three fusion stages. The feature channel changes after using CFECA are shown in Fig. 5.5.

It is observed that after the ECA module is used directly, the model performance in all fusion stages has slightly decreased. This is because in the fusion process, the original fusion method simply concatenates the data of the two modalities together.

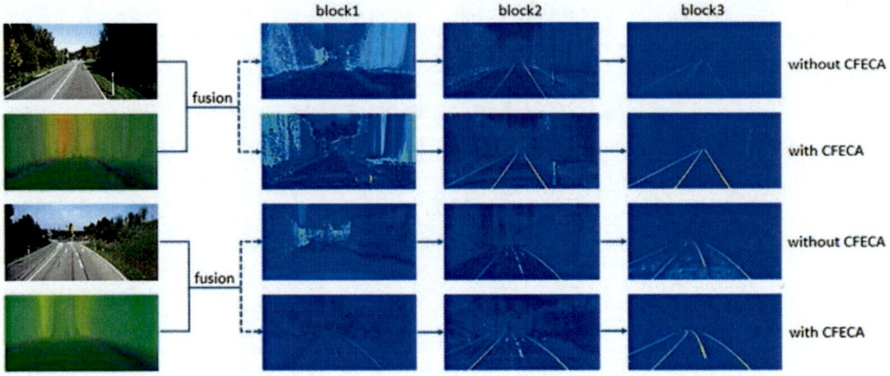

Fig. 5.5 Visualization results of the fusion with and without CFECA in the feature channels of three different convolutional blocks in early fusion

Using the ECA module after that, all the feature channels of the two modalities are compressed to a 1*1*C vector, and then the information of the adjacent k channels is used to calculate the weight of the channel. In this process, only the $k/2$ channels of the splicing position can use the data information of another modal, and the weight calculation of most feature channels only comes from the information of the adjacent feature channels of the original modal. In contrast, the CFECA module first cross-splices the data of the two modalities. When calculating the weight of the feature channel, it is not only guided by the nearby feature channels of the original modal, but it is also led by the feature channels of the same position of another modal data. In addition, we have empirically found that the k value represents the number of guidance channels used. After cross-splicing, the number of guidance channels from the original modal is reduced by half. It is worth noting that the number of guidance channels used by our method is $2k - 1$, thus ensuring that the information needed to calculate the weights of the feature channels from the original modal is not lost, and the complementarity of the multimodal data is also fully utilized.

It can be observed from Table 5.3 that after using CFECA, the precision of models in all fusion stages has been significantly improved, among which the precision of early fusion increased by 4.09%, which shows that our method has a significant improvement effect on multimodal fusion. The calculation of additional feature channel weights will ultimately and inevitably lead to an increase in inference time. However, the FPS of all fusion stages only slightly decreases after using CFECA, which shows that our method does not increase too much computational overhead.

As shown in Table 5.4, we show some qualitative results of our algorithm in early fusion, and the corresponding results of Mid-1 fusion and Mid-2 fusion are shown in Table 5.5 and 5.6. We compared early fusion with our method using ResNet-18 and ResNet-34 pre-training, respectively. When using ResNet-34 for pre-training, our model performed the best, with the highest precision of 56.03% and the best F2 score of 80.75%. It is observed that after pre-training with ResNet-18, the

Table 5.4 The performance of using different pre-trained parameters in early fusion. Here, we use "CFECA" presents ResNet-34 without pretrained parameters, and "CFECA-R-34" presents ResNet-34 with pretrained parameters, and the same abbreviation is used in the following sections. The top row in the table denotes early fusion without any strategy

Algorithm	REC	PRE	F1	F2	Acc	FP	FN
–	92.78	50.28	64.46	78.30	98.59	1.31	0.10
R-18	92.23	51.38	65.25	78.57	98.65	1.24	0.11
R-34	93.16	50.44	64.65	78.63	98.58	1.33	0.09
ECA	93.42	48.45	63.00	77.63	98.49	1.42	0.09
ECA-R-18	93.29	49.60	63.79	77.9	98.53	1.38	0.10
ECA-R-34	92.35	50.17	64.11	77.79	98.56	1.33	0.11
CFECA	92.40	54.37	67.70	80.05	98.80	1.10	0.10
CFECA-R-18	92.16	54.66	67.91	80.12	98.82	1.08	0.11
CFECA-R-34	92.82	56.03	68.97	80.75	98.93	0.99	0.09

Table 5.5 Performance of different pre-trained strategies in Mid-1 fusion

Algorithm	REC	PRE	F1	F2	Acc	FP	FN
–	92.19	50.56	65.27	78.49	98.65	1.25	0.11
R-18	92.19	51.97	65.74	78.83	98.69	1.20	0.11
R-34	92.24	51.85	65.73	78.97	98.71	1.21	0.11
ECA	91.08	50.80	64.28	78.09	98.47	1.27	0.12
ECA-R-18	91.91	52.04	65.61	78.57	98.67	1.22	0.11
ECA-R-34	92.28	50.85	65.66	78.83	98.69	1.20	0.11
CFECA	92.16	52.21	66.08	79.35	98.71	1.19	0.10
CFECA-R-18	91.80	53.76	66.86	79.53	98.78	1.09	0.11
CFECA-R-34	92.86	54.77	67.93	80.43	98.89	1.03	0.09

Table 5.6 Performance of different pre-trained strategies in Mid-2 fusion

Algorithm	REC	PRE	F1	F2	Acc	FP	FN
–	91.82	50.20	63.88	77.72	98.34	1.33	0.12
R-18	91.64	50.75	64.19	78.01	98.42	1.28	0.12
R-34	91.66	50.68	64.17	77.97	98.40	1.29	0.12
ECA	91.68	49.92	63.75	77.34	98.28	1.25	0.11
ECA-R-18	91.86	50.66	64.53	77.94	98.63	1.26	0.11
ECA-R-34	92.02	50.05	64.40	77.92	98.39	1.27	0.12
CFECA	90.62	52.19	65.70	78.06	98.72	1.15	0.13
CFECA-R-18	91.82	50.66	64.53	77.94	98.60	1.26	0.11
CFECA-R-34	92.41	53.11	66.71	79.59	98.73	1.17	0.11

performance of the model hardly changes, but the performance of the model has improved to a certain extent after pre-training with ResNet-34. Our method also got the lowest FP, which shows that our model works better for tasks like separating lane lines.

Table 5.7 Performance of different algorithms on KITTI-aug testing set. Here, our models using CFECA are all carried out in the early fusion stage

Algorithm	Size(M)	REC	PRE	F2	Acc	FPS
LaneNet	285.7	80.97	32.81	60.97	96.87	69.1
SCNN	270.3	88.61	30.37	63.06	97.07	14.4
ENet-SAD	11.0	91.21	33.96	66.90	97.44	22.5
CFECA	17.0	92.40	54.37	80.05	98.80	61.1
CFECA-R-18	17.0	92.16	54.66	80.12	98.82	56.2
CFECA-R-34	25.0	92.82	56.03	80.75	98.93	58.7

We also compared our models (early fusion with different pre-trained ResNet models) with advanced models SCNN, LaneNet, and ENet-SAD on the KITTI-aug dataset. SCNN and LaneNet load pre-trained VGG-16 weights to speed up learning. To be fair, we trained SCNN and LaneNet for 10,000 iterations (equivalent to 175 epochs); they stopped optimization after 5000 iterations. ENet-SAD is trained from scratch, and we trained it for 30,000 iterations (add SAD from 20,000 iterations). Our models trained 200 epochs, and they almost converged after about 150 epochs. The results are shown in Table 5.7 and Fig. 5.6. Our model performs better in the test set.

Moreover, to better illustrate the effectiveness of the proposed method, we embed a variety of attention modules into the fusion for comparison. Table 5.8 shows the comparison of the experimental results. Our model has the highest precision and F2-measure in the test, and the FPS is also at an average level. It is worth noting that although A2Net achieved a high FPS in the test, 36G of GPU memory is required to train the model, and 15G of GPU memory is required to train CCNet. We performed multi-GPU training for these two models separately (our model requires no more than 8G GPU memory; others need about 6G GPU memory). In addition to requiring real-time and high-precision capabilities, self-driving vehicle models must also aim for lightweight design. In the test, all models except A2Net and CCNet require less than 2G memory on GPU. A2Net and CCNet require 4G and 3.3G of memory on the GPU, respectively, and our model achieves the highest accuracy while maintaining the lightest weight.

It can be seen that when the attention mechanism module is directly embedded, only the F2-measure of CCNet has a weak advantage over the baseline. The A2Net model in the spatial attention mechanism has the largest size, but the F2-measure is also the lowest. This may be due to the spatial attention mechanism often calculating the weight of each position based on the global description, but the curved lane line itself has the characteristics of being long and narrow, and excessive spatial attention to the global features may cause interference. It can be seen that our model can reach at least 58.7 FPS with the best performance.

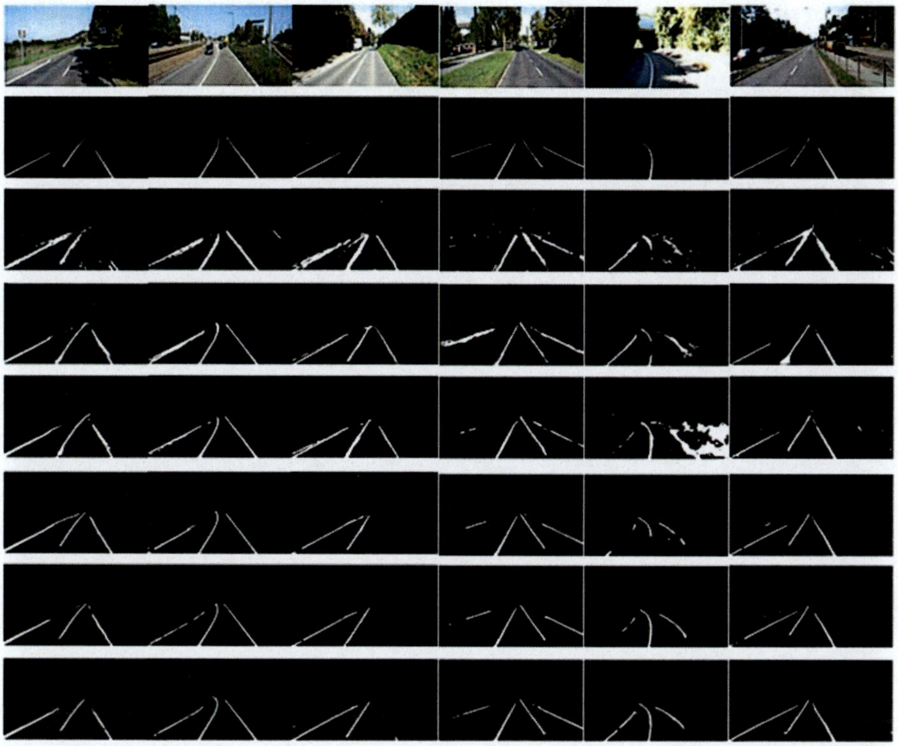

Fig. 5.6 Examples of the test results on KITTI-aug dataset: the rows from top to bottom are input images, ground truth, and output from LaneNet, SCANN, ENet-SAD, CFECA, CFECA-R-18, and CFECA-R-34. All models using CFECA are carried out in the early fusion stage

Table 5.8 Performance of different attention modules in early fusion. Among them, SENet and SKNet are channel attention mechanisms, A2Net is a spatial attention mechanism, CCNet is a self-attention mechanisms, and BAM and CBAM are hybrid domain attention mechanisms

Module	Size(M)	REC	PRE	F2	FP	FN	FPS
Baseline	25.0	93.16	50.44	78.63	1.33	0.09	80.4
SENet	25.0	92.12	51.99	78.45	1.24	0.11	61.8
SKNet	25.1	90.93	51.50	77.49	1.25	0.13	58.1
A2Net [6]	25.3	92.36	45.01	75.17	1.56	0.11	69.6
CCNet [12]	25.1	92.12	52.68	78.87	1.20	0.11	47.5
BAM [15]	25.1	92.50	51.12	78.23	1.30	0.11	58.4
CBAM [21]	25.0	90.98	51.69	77.57	1.24	0.13	64.6
ECA	25.1	92.35	50.17	77.79	1.33	0.11	78.3
CFECA (ours)	25.0	92.82	56.03	80.75	0.99	0.09	58.7

5.4 Adaptive Strategies in Multimodal Fusion Segmentation

In this part, we propose a novel adaptive multimodal fusion network, MIMF, that is driven by the mutual information between the input data and the target recognition pattern. Due to the varied weather and road conditions, the real scenes can be far more complicated than those in the training dataset. That constructs a non-ignorable challenge for multimodal fusion models that obey fixed fusion modes, especially for autonomous driving. To address the problem, we leverage mutual information for adaptive modal selection in fusion, which measures the relationship between the input and target output. We therefore designed a weight-fusion module based on MI and integrated it into our feature fusion lane line segmentation network. We evaluate it with the KITTI and A2D2 datasets, in which we simulate the extreme malfunction of sensors like the modality loss problem. The result demonstrates the benefit of our method in practical application and informs future research into the development of multimodal fusion as well.

5.4.1 MIMF Network

In this part, we select the common middle feature fusion (MF) as the backbone network, which presents robustness in general tests and is regarded as the balance among early fusion, middle fusion, and late fusion [16]. We use an encoder-decoder architecture. In this way, the network can be easily modified and compared to the performance change. The network comprises two pipelines in the encoder for point clouds and images, with three convolutional blocks in both branches. Except for the first, we replace the convolutional blocks with ResNet-34 blocks to process more complex features in the images. We fuse the features of two modalities by concatenation when two pipelines merge, as shown in Eq. (5.7). The information will be mixed up in the following convolutional layers. Each convolutional block includes a convolution layer, a batch normalization layer, and a ReLU activation layer. The blocks in the decoder distinguish the ones in the encoder because they use transposed convolution to recover the feature maps. To make better use of the raw data, we add skip connection between the encoder and decoder layers. In MF, we do not assign a fusion weight; instead, the network learns the adaptive weight. But in MIMF, we embed the DIM module to provide a prior weight that can not only work as regularization but also avoid the influence of bad observations.

$$\boldsymbol{Z}(\boldsymbol{X}) = \boldsymbol{A}\boldsymbol{W}_0\left[\boldsymbol{X}_1, \boldsymbol{X}_2\right]^{\mathrm{T}} = \boldsymbol{W}_0\left[\alpha_1\boldsymbol{X}_1, \alpha_2\boldsymbol{X}_2\right]^{\mathrm{T}} \tag{5.7}$$

DIM was proposed by Hjelm et al. [19] based on MINE [4], which is regarded as an efficient estimator for mutual information between two feature maps in neural networks. In this chapter, we modify the DIM to fit our fusion network. For two variables X and Y, their mutual information $I(\boldsymbol{X}; \boldsymbol{Y})$ is:

$$I(\boldsymbol{X}; \boldsymbol{Y}) = \Sigma_{x \in \boldsymbol{X}} \Sigma_{y \in \boldsymbol{Y}} p(x, y) \log \frac{p(x, y)}{p(x)p(y)} \tag{5.8}$$

$$I(\boldsymbol{X}; \boldsymbol{Y}) = D_{KL}\left(P_{XY} \| P_X \otimes P_Y\right) \tag{5.9}$$

where D_{KL} is the KL divergence. It is defined as:

$$D_{KL}(P \| Q) := E_P\left[\log \frac{\mathrm{d}P}{\mathrm{d}Q}\right] \tag{5.10}$$

We mark P_{XY} as J and $P_X \otimes P_Y$ as M. By using the DV distribution form and the nature of KL divergence, we obtain the lower bound $\hat{I}$ of $I(X; Y)$ as below:

$$I(\boldsymbol{X}; \boldsymbol{Y}) \geq \hat{I}(\boldsymbol{X}; \boldsymbol{Y}) = E_J\left[T_\omega(x, y)\right] - \log E_M\left[e^{T_\omega(x, y)}\right] \tag{5.11}$$

where $T_\omega : \mathrm{x} \times \mathrm{y} \rightarrow \mathbb{R}$ is a function parameterized by that can be used in Eq. (5.11) to approximate $I(\boldsymbol{X}; \boldsymbol{Y})$. We simply present a sample of T_ω and consider it enough to the expected function [20]. The multimodal features $.X1$, $.X2$ are assumed to have the same dimension, which can be achieved by feature alignment, and the size is (C, H, M, N). C is the number of channels in convolutional layers, and (H, M, N) is the size of a channel. We note the map in each channel as $\boldsymbol{X}_{i_n}, n \in$ [1, C]. Therefore, we rewrite Eq. (5.11) as:

$$I\left(\boldsymbol{X}_{i_n}; Y\right) \doteq \hat{I}\left(\boldsymbol{X}_{i_n}; \boldsymbol{Y}\right) = \log\left(\Sigma S e^{u - u_{\max}}\right) + u_{\max} - \log(\Sigma S) - \frac{\Sigma\left(u_{avg} \cdot S\right)}{\Sigma S} \tag{5.12}$$

where $S = 1 - \bar{S}$ and $\bar{S} = H \times H$ is a diagonal matrix. Besides, $U = X_n \times Y$, and $u_{\max}$ is the maximum value in matrix, while u_{avg} is the average value. Therefore, the mutual information is represented as below $C \in \{1, 2\}$ in our model:

$$I\left(\boldsymbol{X}_i; Y\right) = \frac{1}{C} \sum_{n=1}^{C} I\left(\boldsymbol{X}_{i_n}; \boldsymbol{Y}\right) \tag{5.13}$$

(A) Recurrent Training Process

We present how to compute the MI. To apply it in the weighted fusion model, we integrate it into the network and training-testing procedure as well. As illustrated in Fig.5.7, we take two branches from the DIM block, which are for two data. The DIM block computes the mutual information between them and the expected feature Y, respectively. However, we cannot obtain Y ahead of the network computation.

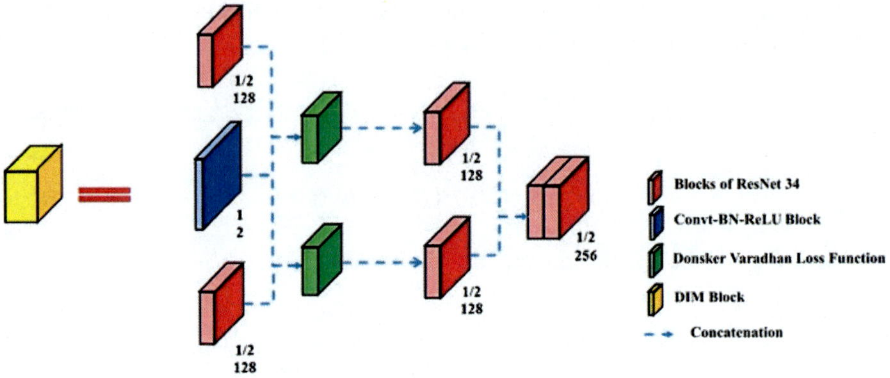

Fig. 5.7 The structure of the DIM block in our fusion network. It adopts features from two modalities, images and point clouds, respectively, as $\boldsymbol{X}_1$ and $\boldsymbol{X}_2$, while taking the features from the last convolution layer as $\widehat{\boldsymbol{Y}}$. Then, DIM block figures out the mutual information $I\left(\boldsymbol{X}_i, \widehat{\boldsymbol{Y}}\right)$

Instead, we make the time-continuity assumption: for each i, $\boldsymbol{X}_i$ is a given stable time sequence, which means $\boldsymbol{X}_i^t \approx \boldsymbol{X}_i^{t-1}$ and $\boldsymbol{Y}^t \approx \boldsymbol{Y}^{t-1}$. In autonomous driving, that indicates two frames of a sensor's observations are similar because of the continuity of scenarios and events. With this assumption, when we acquire a well-trained model in a test, we can treat the recognition of the last frame as an approximation of the target at the current time, especially in a sequence model. Obviously, the fault rate of the last recognition will be increased. But when DIM is integrated into a robust backbone, we can ignore it most of the time and use a reset strategy to reduce the cumulative error. However, we have only implemented a single-frame recognition model and lack enough time series data, so we simply compute the current data cyclically. Specifically, we compute on it for the first time to simulate the "last frame result" and use it in the second computation. Therefore, we finish the DIM process approximately in testing. Figure 5.8 presents the overall structure of MIMF. The yellow block is the DIM module, and the rest is the MF baseline. The RGB images and point clouds are processed in two separate pipelines in the encoder and get fused in the DIM module. The sizes of feature maps are not changed in DIM. That means DIM is flexible for most models. When DIM outputs $I\left(\boldsymbol{X}_i, \widehat{\boldsymbol{Y}}\right)$ as above, we normalize them by:

$$\alpha_i = \frac{I\left(\boldsymbol{X}_i, \widehat{\boldsymbol{Y}}\right)}{\sum I\left(\boldsymbol{X}_i, \widehat{\boldsymbol{Y}}\right)} \tag{5.14}$$

(B) MI as Fusion and Regularization

With Eq. (5.14), we obtain the fusion weight A in Eq. (5.7). As a prior knowledge of the target of the tasks, MIMF pre-fetches data with a bias. It further forces the fusion models to focus on more relevant information in testing. In practice, the bias causes it to be unaffected by fault measurements or information loss data in complex

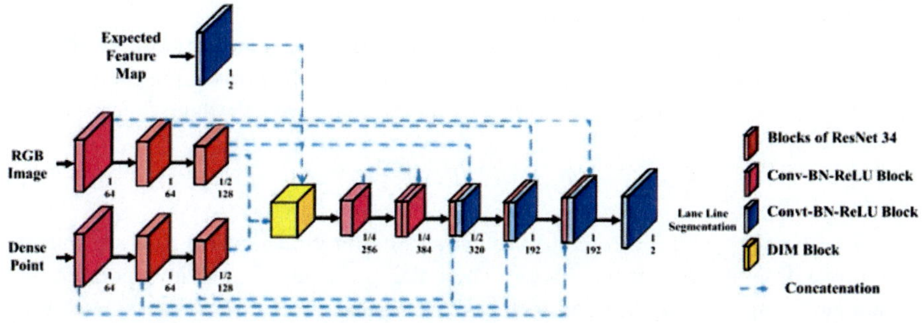

Fig. 5.8 The overview of the architecture of MIMF. It consists of a standard feature fusion in the middle of an encoder-decoder network and a DIM block during fusion. Before fusion, MIMF had two individual pipelines to process different modalities of data

scenes. In addition, we observe that MIMF performs better on the normal data. The random regularization effect, which is similar to the dropout, is used to explain the result. MI will be treated as a random process with a distribution different from that of the noised data because it is independent of the network or data and is determined by both the data and the target at the same time. Therefore, by learning the data-independent input, the network avoids overfitting the data. Notice that our method can only operate in the case when at least one modality has good observation. Otherwise, the dominated data will lead to serious problems.

5.4.2 *Experiment*

(A) Dataset and Metrics

To evaluate our models, we select pictures by ignoring the roads with intersections or without forward lines. Finally, we pick up around 400 data pairs from the KITTI road detection track [10] and around 1000 pairs from the A2D2 dataset [11]. We use 60% of the data as the training set, 10% for validation and the rest for testing. The image resolution is in KITTI and in A2D2. KITTI uses a 64-line Velodyne to generate point clouds, but A2D2 combines one 8-line and two 16-line LiDARs. The difference in LiDARs causes the gap in performance, but it would not matter in the evaluation of the adaptive fusion. Because the KITTI dataset has no lane line labels, we add pixel-level annotation to it by hand. Labeled lines are supposed to be parallel not only to the driving direction but also on the driving area. To reduce noise in the annotation, we do not estimate any markings behind obstacles like vehicles and poles on the roadside. Different from KITTI, A2D2 provides similar lane line labels, but they ignore the intervals in the dash lines.

We focus more on the recall of lane lines and compute it as the lane accuracy. We also consider the F2 score to be balanced in case the network overfits any class, and we count the mean recall in both classes as the mAcc.

(B) Implementation and Training

To integrate LiDAR point clouds and RGB images into the same network, projection and value normalization are essential in preprocessing. To project the point clouds onto the image plane, given a point $P_v = (x_v; y_v; z_v)^T$, we calculate:

$$P_v = K_v [R_v; T_v] P_v \tag{5.15}$$

where Kv, Rv, and Tv refer to the camera calibration matrix, rotation matrix, and translation matrix, respectively. Then the projected front-view point cloud reflectance map will be cropped to the same size as the RGB images. After that, the value of both the reflectance map and the RGB images will be normalized to the [0,1] interval. Following data preprocessing, we train our model on 2 datasets for 250 epochs each. As shown in Sect. 5.4.1, we generate the simulated ground truth of target features in the first round of training, in which we also get the result of the original MF model. Then, in the second round, we train the MIMF with the features. In testing, we use the pre-trained MF, just as in training, to get target features and test MF and MIMF.

(C) Result and Analysis

We present the training record in Fig. 5.9 and the result of testing in Fig. 5.10. Note that we only put the training record of the last 50 epochs in the figures, in which we can see the MIMF performs worse than MF at first, but they converge together at last. That indicates the random disturbance from the independent mutual information. However, in testing, we observe that MIMF performs better than MF by 1 2%, which indicates the potential regularization function of MI-driven models on the small training data. Note that due to the unstable calculation in MIMF, we have different outputs in testing, for which we process it tenth and count the average value. Though

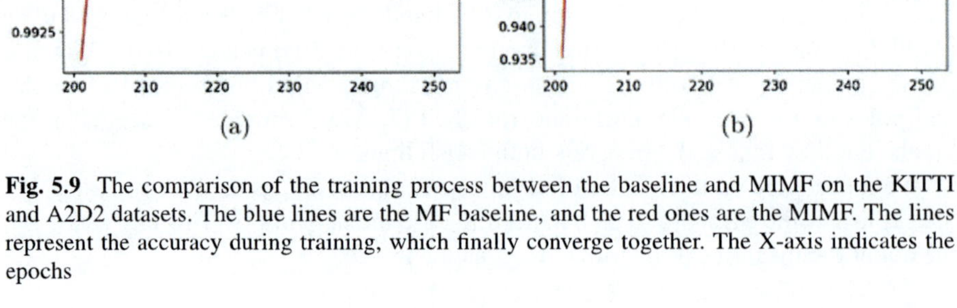

Fig. 5.9 The comparison of the training process between the baseline and MIMF on the KITTI and A2D2 datasets. The blue lines are the MF baseline, and the red ones are the MIMF. The lines represent the accuracy during training, which finally converge together. The X-axis indicates the epochs

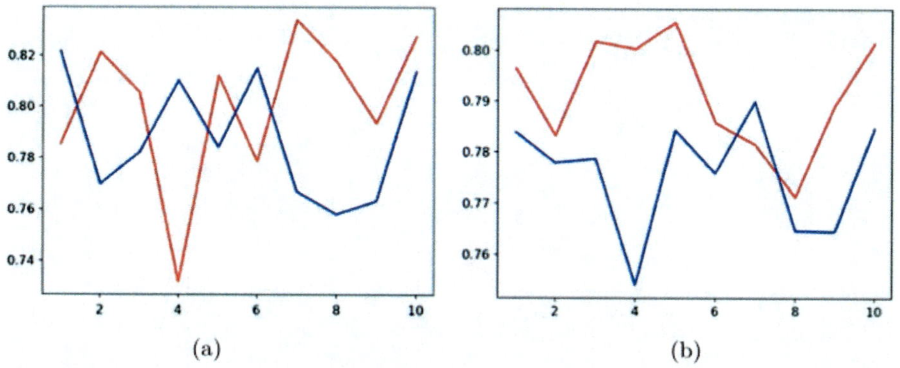

Fig. 5.10 The comparison of the testing process between the baseline and MIMF on the testing datasets. The blue lines are the MF baseline, and the red ones are the MIMF. The X-axis indicates the epochs

the result is not deterministic, the DIM in MIMF outputs the stable fusion weight, which is 1.25:0.75 for image and point cloud fusion in normal data in KITTI. For A2D2, the ratio is 1.35:0.65, and that matches the prior in the dataset when we declare that the LiDARs in A2D2 are not suitable for segmentation tasks. We further complete the modality loss on the KITTI dataset. MIMF can keep the performance elimination in an acceptable range with prior knowledge of the MI of each sensor, whereas MF only recalls 50.29% of the lane pixels, which is far less than the result on normal data.

5.5 Video Multimodal Fusion Segmentation

Multimodal data fusion is becoming a trend in the field of autonomous driving, especially for lane detection. In the process of driving, sensors often encounter problems such as modality imbalance, changing illumination, and so on. Therefore, it is worthwhile to study the problems of applying multimodal fusion for lane detection and modality imbalance in the fusion process. In this chapter, we propose a novel multimodal model for lane detection in which an attention mechanism is embedded into the network to balance multimodal feature fusion and improve detection capability. In addition, we use multi-frame input and long short-term memory (LSTM) networks to solve the shadow interference, vehicle occlusion, and mark degradation problems. At the same time, the network can be applied to the task of lane detection. In order to verify the effect of multimodal application and attention mechanism on fusion, we have designed adequate experiments on the processed continuous scene KITTI dataset. The results show that precision increases by about 15% when LiDAR is added compared with RGB only. Besides, the attention mechanism obviously improves the performance of multimodal detection by balancing multimodal features (Fig. 5.11).

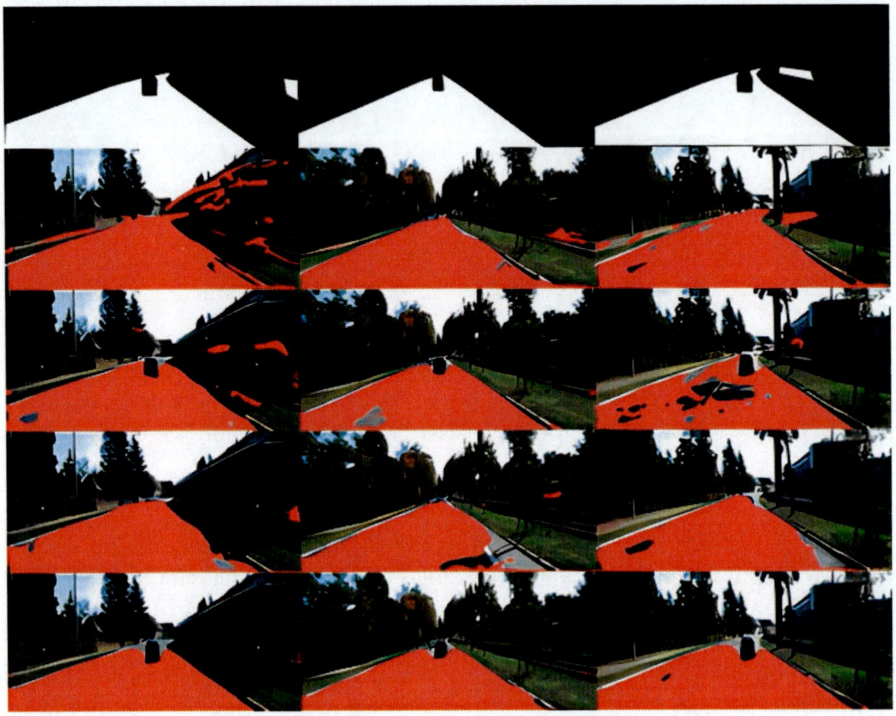

Fig. 5.11 Lane detection in continuous scenes. Corresponding ground truth annotations in the top row. Second row: lane detection achieved with a RGB-only network. The RGB+ LiDAR network detected a lane in the third row. Fourth row: lane detection obtained by a multimodality network with SKNet attention. Bottom row: lane detection obtained by a multimodality network with TripletNet attention

5.5.1 *Method*

In this section, we first explain the significance of real-time multimodal fusion in realistic autonomous driving and introduce the attention mechanism to balance fusion features. Following that, we will present the architecture and theory explanation of our method in detail.

(A) Overview

Multimodal fusion can obtain more information, but we consider the modality imbalance between different modes will affect the fusion results. In view of this, we try to add an attention mechanism after the fusion process to balance multimodal features. At the same time, attention mechanisms play a significant role in the use of global information. It can select the region of interest in the feature map, optimize the feature map globally, highlight the key regions in the image, and suppress the interference of non-important regions (background) on the detection effect. In order to verify our idea, we chose SKNet and TripletNet as attention modules to join the

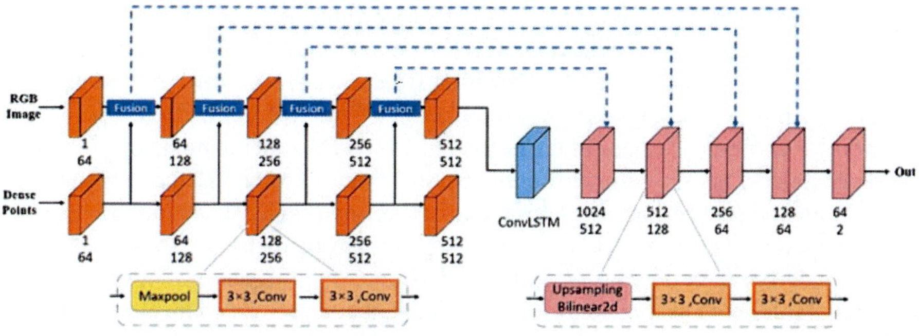

Fig. 5.12 The architecture of the proposed network

network. Among them, the SKNet attention module is inserted after fusion, while the TripletNet attention module is added to each decoder and encoder block. In the following sections, we build our network and describe each step of the network in detail. Figure 5.12 shows the network architecture we proposed.

(B) Network Architecture

1) Encoder-decoder network. One of the most remarkable features of the encoder-decoder network is that it is an end-to-end structure. We regard the lane detection task as semantic segmentation using an encoder-decoder network. The encoder uses convolution and pooling operations to abstract images and extract features, while the decoder recalls and applies the information of targets using deconvolution and up-sampling. Referring to the work of [24], we chose U-Net as the encoder-decoder architecture of the model.

In the U-Net, the blocks are classical full convolution networks. We input LiDARs and RGB images into U-Net at the same time, and then they pass through five consecutive convolution blocks. Each block contains two convolution layers of the same size. A pooling operation is added between each convolution block to down-sample the feature map. The number of channels is doubled each time, and the size of the feature map is reduced by two times. It is worth mentioning that for subsequent inputs to the LSTM network, the number of convolution kernels in the last convolution block has not changed relative to the previous one. In the decoder blocks, deconvolution and up-sampling correspond to operations in the encoder. Skip connection is applied to the encoder and decoder by concatenating, which doubles the size of the feature map.

Figure 5.12 shows the detailed network architecture of encoder-decoder blocks. The encoder contains two branches, each with the same operation. After input, it first goes through the maxpool layer and then through the double 3×3 convolution operation. Finally, the fusion is performed, and the specific operations of the fusion are shown in the next section. It is worth noting that only the RGB branches remain after the fusion and the LiDAR branches are input directly into the next block. In the decoder block, the blue dashed line indicates a skip connection from the encoder. The decoder consists of a single branch, and the basic structure of a block is similar to those in the encoder. However, in the decoder, the first operation of the block is up-sampling instead of max-pooling. The output from the block with the same number of layers is then added to the output from the encoder until the last layer.

2) Fusion network. Previously, we introduced four fusion methods. The fusion method we chose is a variant of cross-fusion, which fuses the features of each layer in the encoder. This strategy allows multimodal features to be fused at any network depth, rather than limiting them to a single level as previously mentioned. In the fusion method, we do not choose a two-way cross but focus on a one-way cross on the image branch. Because it is multiple frame input, it can not only make the network lighter but also prepare for our subsequent attention mechanism. In addition to that, we employ the concatenate operation instead of the add operation. In the fusion stage, the input tensors at depth j are denoted as I_j^r and I_j^d. We feed them to layers L_j^r and L_j^d, respectively. The detailed expression is as follows:

$$I_j^r = \mathrm{Cat}\left(\left(L_{j-1}^r\right)\left(p_{j-1}L_{j-1}^d\right)\right) \tag{5.16}$$

or

$$I_j^r = L_{j-1}^r + p_{j-1}L_{j-1}^d \tag{5.17}$$

where $p_j \in$ R and j $\in \{1, 2, 3, 4\}$ are trainable cross-fusion parameters and Cat indicates the concatenation operation. During training, these fusion parameters are automatically modified to integrate the two information modes. In SKNet, we choose 5.16 in order to adapt to the input of attention mechanism network, and in TripletNet attention, we choose the common fusion method 5.17.

(C) Attention Mechanism Network

For end-to-end network, channel attention mechanism has the characteristics of a small amount of calculation and light weight. The SKNet attention and TripletNet attention are lightweight and modular, which can be embedded into a network. They are also efficient and widely used. In SKNet, selective kernels are added to select different spatial scales of information through an inter-channel soft attention mechanism. The architecture of the selective kernel (SK) unit is shown in Fig. 5.13. SK convolution consists of three operations: Split, Fuse, and Select. The Split operation produces multiple paths with various convolution nucleus sizes, corresponding to different receptor field sizes of neurons. As shown in Fig. 5.13, the output is obtained by convoluting X with Kernel 3*3 and Kernel 5*5 separately. Fuse combines and aggregates information from two branches to obtain global representations of selective weights. Then the results from the two branches are fused through element-wise summation.

$$\mathrm{U} = \widehat{U} + \widetilde{U} \tag{5.18}$$

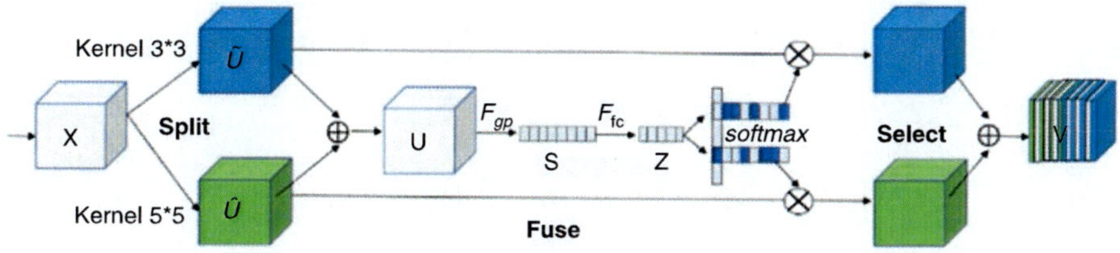

Fig. 5.13 The architecture of the selective kernel (SK) unit

F_{gp} represents the global average pooling (GAP) operation, and F_{fc} is a two-level full connection layer that reduces and then increases dimensions.

The Select operation aggregates feature mappings of convolution cores of different sizes based on the selection weight. Select operation weights $\widehat{U}$ and $\widetilde{U}$ using two weight matrices, a and b, and then sum them to get the final output vector V:

$$V_c = a_c\widehat{U}_c + b_c\widetilde{U}_c, a_c + b_c = 1 \tag{5.19}$$

where $\mathrm{V} = [V_1, V_2, \ldots, V_c]$, $V_c \in R^{\mathrm{H}\times\mathrm{W}}$. $a_c + b_c = 1$ in case of two branches.

In the TripletNet module, for the input tensor, the TripletNet attention establishes the dependency relationship between dimensions through rotation operation and residual transformation and encodes the inter-channel and spatial feature information with low computational overhead. The TripletNet attention network structure has three branches. Given an input tensor $\chi \in R^{\mathrm{C}\times\mathrm{H}\times\mathrm{W}}$, the first branch is the channel attention calculation branch, where interactions are established between the H dimension and the C dimension. In the second branch, interactions are established between the C dimension and the W dimension, and in the third branch, interactions are established between the H dimension and the W dimension to construct spatial attention. Finally, Avg is calculated by adding the output features of the three branches. Detailed algorithm procedures can be found in [23].

4) LSTM network. LSTM models multiple consecutive frames of a driving scene as a time series and accepts as input the feature map extracted by the encoder CNN on each frame. The convolutional LSTM (ConvLSTM), which not only has the time series modeling ability of LSTM but can also depict local characteristics like CNN, was proposed by X. Shi et al. In addition, the authors have experimentally demonstrated that ConvLSTM is more effective than LSTM in acquiring spatiotemporal relationships. ConvLSTM is essentially the same as LSTM, where the output of the previous layers is fed as the input of the subsequent layer. However, ConvLSTM replaces feedforward calculation of LSTM gates with a convolution operation, which saves much time and computational cost. The working principle of ConvLSTM can be expressed by the following formula:

$$i_t = \sigma\left(W_{xi} * X_t + W_{hi} * H_{t-1} + W_{ci}^{\circ} C_{t-1} + b_i\right) \tag{5.20}$$

$$f_t = \sigma\left(W_{xf} * X_t + W_{hf} * H_{t-1} + W_{cf}^{\circ} C_{t-1} + b_f\right) \quad (5.21)$$

$$C_t = f_t \circ C_{t-1} + i_t \circ \tanh\left(W_{xc} * X_t + W_{hc} * H_{t-1} + b_c\right) \quad (5.22)$$

$$o_t = \sigma\left(W_{xo} * X_t + W_{ho} * H_{t-1} + W_{co}^{\circ} C_t + b_o\right) \quad (5.23)$$

$$o_t = \sigma\left(W_{xo} * X_t + W_{ho} * H_{t-1} + W_{co}^{\circ} C_t + b_o\right) \quad (5.24)$$

where "*" denotes the convolution operator and "∘," as before, denotes the Hadamard product. The design is that all the inputs $X_1, \ldots, X_t$; cell outputs $C_1, \ldots, C_t$; hidden states $H_1, \ldots, H_t$; and gates i_t, f_t, and o_t of the ConvLSTM are 3D tensors whose last two dimensions are spatial dimensions.

5.5.2 *Experiments*

For experiments designed to verify the effectiveness and importance of attention mechanisms in multimodality fusion, we chose two popular attention mechanisms: SKNet attention and TripletNet attention. The visual comparison results are illustrated in Fig. 5.11. We also do ablation experiments to show why multimodal feature fusion is better than single mode.

(A) Training Procedure

During the training, the sizes of images are sampled to 512×256, and the batch size is 2. We use the SGD optimization algorithm in the network and set the learning rate as 0.01. The number of epochs is 60, and the network is trained in an end-to-end way. The fusion networks were run by an NVIDIA GTX 1080 GPU and implemented in PyTorch. The weights of the network are saved and evaluated at each epoch. The optimal evaluation weight will be saved and tested in the last test.

(B) Metrics

The evaluation metrics usually include precision (PRE) and recall (REC). In addition, we also add F-measure as an important evaluation index so as to comprehensively reflect the performance of lane detection. F-measure is obtained from the precision rate and recall rate of pixel-level evaluation, and we usually use F1-measure (= 1). The formulas are shown as follows:

$$\text{PRE} = \frac{TP}{TP + FP} \quad (5.25)$$

$$\text{REC} = \frac{TP}{TP + FN} \quad (5.26)$$

$$\text{F1} = 2\frac{PRE \times REC}{PRE + REC} \quad (5.27)$$

We set the lanes in the image as a positive class and the others as negative classes in the task of lane detection, which can be regarded as a binary classification way in essence. In Formulas (5.26) and (5.27), TP indicates the number of lane pixels truly predicted as lane, FP indicates the number of background pixels incorrectly predicted as lane, and FN indicates the number of lane pixels incorrectly predicted as background.

(C) Performance Analysis

The test results, which are on the KITTI dataset, are displayed in Fig. 5.11. There are a total of six lines of results, which are introduced in the figure. It can be known that single mode (RGB) has the poorest detection performance. In this mode, the detection effect is often inaccurate due to the lack of some necessary information, such as depth information. Some areas similar to the lanes are often detected, resulting in large background interference. At the same time, the influence of some light conditions will also affect the detection effect. Some shadows are likely to be ignored as non-lane areas. Therefore, it is not enough to use a single mode. Our model uses a multimodal fusion strategy, as shown in the fourth line of the figure; the detection effect is greatly improved compared with the single mode. The redundancy has been greatly reduced. This improvement mainly comes from the addition of LiDAR information. The network can compensate for the limitation of using a single mode by using multimodal fusion to improve detection performance.

However, this effect is not what we expected. Although there is a greater improvement when compared to single mode, the detection effect still has many flaws. As shown in the figure, there are some missing holes in some detection areas, while there is still a small amount of background interference. We consider that this is an imbalance between multimodal features. Because the same scene is often distinct in different modes, the information mismatch may result in some empty areas in the effect map. Therefore, we introduce an attention mechanism to promote the fusion of multimodal features, as shown in the last two lines of the figure. It can be seen that background interference has barely occurred and that the lane area can be completely detected. The SKNet attention has better resilience to scale changes and is able to take advantage of global information to detect areas that are not easily detectable, such as opposite lanes. After a deeper analysis, SKNet activates cortical neurons to dynamically change their own receptive fields on the basis of different stimuli and carries out selective kernel transformation on all convolution kernels >1, fully benefiting from the smaller theoretical parameters and flop dividends resulting from group or depth-wise convolution, thereby increasing the design of multiple and dynamic selection. After the introduction of the attention mechanism, the features of different channels and modes communicate and recombine with each other, so that more significant areas can be detected, so that the network not only has better multi-scale detection ability but also makes full use of multimodal features. The TripletNet attention is excellent for locally detected areas with little background interference or void areas. Because the TripletNet attention has a negligible computational overhead, we incorporate it into each block. It interacts with dimensions through three paralleled branches, of which two branches are used for capturing the cross-

Table 5.9 Performance on different strategies

Network	PRE/%	REC/%	F1/%
RGB only	74.53	**97.11**	84.34
RGB+ LiDAR	88.97	89.38	89.17
Q. Zou et al. [16]	76.76	95.30	85.03
PLARD	92.38	90.64	91.50
Multimodality with SKNet	91.25	80.73	85.67
Multimodality with TripletNet	**93.08**	92.35	**92.71**

Bold values highlight the significant improvement in performance

latitude interaction between space and channel C, while the last one is responsible for constructing spatial attention. The output from these branches is aggregated evenly at the end. In this way, the information between different dimensions can be fully exchanged. In the former encoder, the two mode branches themselves are optimized, which is conducive to the next fusion. After the fusion, the multimodal features can be effectively balanced in the decoder so that the multimodal features can be fully utilized. In addition, it can also optimize the features after skipping connections. In this way, the detection ability of the whole network has been greatly improved.

The specific evaluation indicators are shown in Table 5.9. The first row represents single mode, the second is multi-mode, and the last is multi-mode with attention mechanism. For PRE, using multimodal is about 15% higher than single modal, which shows the importance and superiority of using multimodal fusion effect. Multimodality with attention has a high detection ability with increasing precision. Recall is also an important indicator for evaluating the proposed method. We can see that the recall rate of single mode is the highest. The recall rate is down a lot after fusion. Therefore, in order to solve the imbalance problem of multimodal feature fusion, we introduced an attention mechanism. It can be seen that after the attention network is inserted, the recall rate has greatly improved. For the goal of comprehensive evaluation and considering precision and recall, the F1-measure can be regarded as the correct index to evaluate the experimental method. Multimodal fusion improves F1-measure by nearly 5% compared to single mode. Besides, we can know that the TripletNet attention has achieved the highest F1-measure. The experimental results verify the effectiveness of the attention mechanism for multimodality fusion. Finally, we compare with other lane detection methods, including the most competitive algorithm PLARD using multimodal fusion and the algorithm [24] using multi-frame input. We trained them on the continuous scene dataset we built. The test results show that our algorithm performs better on the dataset, proving that the multimodal attention-guided real-time network is very effective. In addition, we also test the running time of the network processing a single frame. Our proposed network reaches a speed of 0.15 s, while PLARD reaches a speed of 1.36 s.

Table 5.10 Ablation study for model on KITTI dataset

Method			Results on KITTI		
Camera	LiDAR	LSTM TripletNet	PRE/%	REC/%	F1/%
✓			74.53	97.11	84.34
✓	✓		+4.36	−2.80	+1.57
✓	✓	✓	+14.44	−7.73	+4.83
✓	✓	✓✓	+18.55	−4.76	+8.37

(D) Ablation Study

We explore the performance differences of various strategies to demonstrate the significant effect of multimodal fusion and the enhancement of attention mechanisms on fusion. First, we compare the different modes. In Table 5.10, we use the camera only as a benchmark and add LiDAR in order to verify the effectiveness of multimodality. As shown in the table, both the prediction rate and the F1-measure rate have been improved. When LSTM is added, all performances improve greatly. Although the detection speed of the whole network is reduced due to the large amount of LSTM computation, it is worthwhile to achieve a high performance improvement by using time series signals. In addition, we add the TripletNet attention in the third column in order to further highlight the role of attention mechanisms in balancing multimodal fusion. The experimental results show that compared with using a camera only, the prediction rate and F1-measure rate have a greater improvement (PRE + 18.55%, F1 + 8.37%). At the same time, the recall rate also shows a certain improvement compared with multimodal fusion.

5.6 Summary

In this chapter, inspired by the channel attention mechanism, we have proposed an optimized LiDAR-camera fusion network using the CFECA module for the task of lane line segmentation. In summary, this work makes the following major contributions to the general knowledgebase:

- Efficient channel attention is proposed to improve LiDAR-camera fusion in lane line segmentation.
- The CFECA (cross-fusion efficient channel attention) module is designed and applied to improve the fusion method and allow abundant LiDAR-camera fusion information to be used simultaneously across channels.
- A method for determining fusion weights, which are transferable in multimodal fusion, is proposed.

But after the network pulls out the features, the features at the same level might not be the best ones for fusion. Therefore, we have studied adaptive fusion in the fusion network.

Inspired by the mutual information (MI) [6], which measures the relation between two variables, people refer to the amount of information in models [18]. A network is supposed to perform at its best during information acquisition. Therefore, information maximization equals fusion efficiency maximization to some extent. To address efficient usage of MI, some research contributes to the MI estimation in neural network [18]. Based on the previous work Deep InfoMax (DIM) [2], we proposed a novel MI-based data fusion that figures the weight for feature fusion dynamically. The central concept of our work is the real-time calculation of the MI value of multimodal features and recognition targets, which broadens the fusion tendency on them. We build an end-to-end model and examine it on LiDAR-camera fusion lane line segmentation task on the KITTI and A2D2 datasets [12].

Then we propose a novel efficient model for lane detection in which the multimodal feature fusion and LSTM network are used to solve some problems such as vehicle obstruction and mark degradation. We introduce an attention mechanism for multimodal fusion, which can balance multimodal feature fusion and capture salient areas around the vehicle. Through attention guidance, the network can be more focused and prominent. In summary, this work makes the following major contributions to the general knowledgebase:

1) We propose an effective real-time model for lane detection, using a fusion strategy to compensate for the limitations of single-mode detection and applying multi-frame input to solve practical problems such as vehicle obstruction and mark degradation.
2) We introduce an attention mechanism for multimodal fusion, which can balance multimodal feature fusion and capture salient areas around the vehicle. Through attention guidance, the network can be more focused and prominent.
3) Our framework can be trained in an end-to-end manner and has more practicality and advantages in engineering. For example, the proposed network can also be used for lane detection. The experimental results show that our network achieves a better comprehensive effect and verify the importance of the attention mechanism in multimodality fusion.

5.7 Challenges and Prospect

The current approaches to lane line segmentation can be divided into three types: camera-based models, LiDAR-based models, and fusion models. The camera-based methods can meet the high frame rate requirements of driving scenes. Additionally, camera images have a high resolution and efficient array storage structure [5], thus providing abundant information under good illumination and fair weather conditions. However, dramatic changes in light have a considerable impact on the performance of the camera. In contrast to a camera, LiDAR retains abundant 3D information in the driving environment, provides accurate distance measurements [8], and is minimally affected by ambient light conditions. However, LiDAR only provides sparse and irregular point cloud data, which can lead to the existence of many empty voxels [9]. Both camera-based and LiDAR-based methods have distinct advantages and disadvantages. To optimally consider the advantages of different sensors in accurately describing the external environment, it is necessary to fuse the data from different sensors [22]. Multimodal fusion involves combining data from different sensors to perform multi-stage feature fusion at different levels [13]. The fusion model mainly fuses the data of multiple modes and complements the information. Through multimodal fusion, this method can improve the accuracy of lane line segmentation. Notably, the redundancy and fault tolerance of the system and the speed and accuracy of decision-making can be improved [2].

However, although the existing multimodal methods can perform well in most scenes, the fusion strategy may fail seriously in some abnormal scenes [6]. For example, bad weather such as rain and fog will cause obstacles to the camera's work. The sensor itself also contains potential perceptual bias, such as noise point cloud intensity in LiDAR. Besides simultaneous interpreting of these external and internal problems, there is another common but disturbing problem in practice. Data flows from different sensors are not always matched in time due to hardware restrictions. These problems lead to data uncertainty, which expands the model performance gap between the dataset and the actual situation and hinders the application of multimodal fusion methods.

Although, in most tasks, the fusion model using multiple sensors is better than using a single sensor, there are also a series of problems to solve, such as what data to fuse, when to fuse, how to fuse, and so on. The three fusion strategies proposed in this chapter had good results in the experiment, but there are still places to be improved. Further research on fusion will be carried out in the future.

References

1. Ashraf, I., Hur, S., Park, Y.: An investigation of interpolation techniques to generate 2D intensity image from LiDAR data. IEEE Access **5**, 8250–8260 (2017)
2. Asvadi, A., Garrote, L., Premebida, C., Peixoto, P., Nunes, U.: Multimodal vehicle detection: fusing 3D-LiDAR and color camera data. Pattern Recognit. Lett. **115**, 20–29 (2017)
3. Bai, M., Mattyus, G., Homayounfar, N., Wang, S., Lakshmikanth, S.K., Urtasun, R.: Deep multi-sensor lane detection. In: 2018 IEEE/RSJ International Conference on Intelligent Robots and Systems (IROS), pp. 3102–3109. IEEE (2018)
4. Belghazi, M.I., Baratin, A., Rajeshwar, S., Ozair, S., Bengio, Y., Courville, A., Hjelm, D.: Mutual information neural estimation. In: International Conference on Machine Learning, pp. 531–540. PMLR (2018)
5. Caltagirone, L., Bellone, M., Svensson, L., Wahde, M.: LiDAR-camera fusion for road detection using fully convolutional neural networks. Robot. Auton. Syst. **111**, 125–131 (2019)
6. Chen, Y., Kalantidis, Y., Li, J., Yan, S., Feng, J.: A2-nets: double attention networks. arXiv preprint arXiv:1810.11579 (2018)
7. Feng, D., Haaseschuetz, C., Rosenbaum, L., Hertlein, H., Glaeser, C., Timm, F., Wiesbeck, W., Dietmayer, K.: Deep multi-modal object detection and semantic segmentation for autonomous driving: datasets, methods, and challenges. arXiv: Robotics (2019)

8. Feng, M., Zhang, L., Lin, X., Gilani, S.Z., Mian, A.: Point attention network for semantic segmentation of 3D point clouds. arXiv: Computer Vision and Pattern Recognition (2019)
9. Garnett, N., Cohen, R., Peer, T., Lahav, R., Levi, D.: 3D-lanenet: end-to-end 3D multiple lane detection. arXiv: Computer Vision and Pattern Recognition (2018)
10. Geiger, A., Lenz, P., Urtasun, R.: Are we ready for autonomous driving? the kitti vision benchmark suite. In: 2012 IEEE Conference on Computer Vision and Pattern Recognition, pp. 3354–3361. IEEE (2012)
11. Geyer, J., Kassahun, Y., Mahmudi, M., Ricou, X., Durgesh, R., Chung, A.S., Hauswald, L., Pham, V.H., Mühlegg, M., Dorn, S., et al.: A2D2: audi autonomous driving dataset. arXiv preprint arXiv:2004.06320 (2020)
12. Huang, Z., Wang, X., Huang, L., Huang, C., Wei, Y., Liu, W.: Ccnet: criss-cross attention for semantic segmentation. In: Proceedings of the IEEE/CVF International Conference on Computer Vision, pp. 603–612 (2019)
13. Li, Y., Zhang, Z., Cheng, Y., Wang, L., Tan, T.: MAPNet: multi-modal attentive pooling network for RGB-D indoor scene classification. Pattern Recogn. **90**, 436–449 (2019)
14. Nikoohemat, S., Diakite, A.A., Zlatanova, S., Vosselman, G.: Indoor 3D modeling and flexible space subdivision from point clouds. ISPRS Ann. Photogramm. Remote Sens. Spat. Inf. Sci. **4**, 285–292 (2019)
15. Park, J., Woo, S., Lee, J.Y., Kweon, I.S.: Bam: Bottleneck attention module. arXiv preprint arXiv:1807.06514 (2018)
16. Peng, C., Ma, J.: Semantic segmentation using stride spatial pyramid pooling and dual attention decoder. Pattern Recogn. **107**, 107,498 (2020). URL http://www.sciencedirect.com/science/article/pii/S0031320320303010
17. Peng, X., Murphey, Y.L., Liu, R., Li, Y.: Driving maneuver early detection via sequence learning from vehicle signals and video images. Pattern Recogn. **103**, 107,276 (2020)
18. Qian, Y., Dolan, J.M., Yang, M.: DLT-Net: joint detection of drivable areas, lane lines, and traffic objects. IEEE Trans. Intell. Transp. Syst. **21**, 1–10 (2019)
19. Tang, J., Li, S., Liu, P.: A review of lane detection methods based on deep learning. Pattern Recogn. **111**, 107,623 (2021). URL http://www.sciencedirect.com/science/article/pii/S003132032030426X
20. Wang, Q., Wu, B., Zhu, P., Li, P., Zuo, W., Hu, Q.: ECA-Net: efficient channel attention for deep convolutional neural networks. arXiv: Computer Vision and Pattern Recognition (2019)
21. Woo, S., Park, J., Lee, J.Y., Kweon, I.S.: Cbam: convolutional block attention module. In: Proceedings of the European Conference on Computer Vision (ECCV), pp. 3–19 (2018)
22. Xu, X., Li, Y., Wu, G., Luo, J.: Multi-modal deep feature learning for RGB-D object detection. Pattern Recogn. **72**, 300–313 (2017)
23. Zhou, K., Chen, L., Cao, X.: Improving multispectral pedestrian detection by addressing modality imbalance problems. In: European Conference on Computer Vision, pp. 787–803. Springer (2020)
24. Zou, Q., Jiang, H., Dai, Q., Yue, Y., Chen, L., Wang, Q.: Robust lane detection from continuous driving scenes using deep neural networks. IEEE Trans. Veh. Technol. **69**(1), 41–54 (2019)

Chapter 6
Multi-Sensor Fusion Localization

Abstract In real-time positioning problems, some unstable situations often occur, such as GPS signal loss and map drifting. In order to have better localization and map buildings, this chapter proposes a coarse-and-fine hybrid positioning system, which integrates the global information and feature-based simultaneous localization and mapping (GF-SLAM). The system can operate adaptively when external signals are unstable and avoid the cumulative error from local methods. Generally, it applies the feature-based SLAM (F-SLAM) for coarse positioning with particle filter. If available, it fuses the accurate global information by extended Kalman filter (EKF) for precise positioning and revises the deviation in mapping, thereby achieving an effective combination of two positioning modes. In addition, we introduce semantic mapping and lifelong localization approaches to recognize semi-dynamic objects in non-static environments. We also propose a generic framework that can integrate mainstream object detection algorithms with mapping and localization algorithms. The mapping method combines an object detection algorithm and a SLAM algorithm to detect semi-dynamic objects and constructs a semantic map that only contains semi-dynamic objects in the environment. In summary, this chapter proposes methods to resolve self-positioning and map drift when positioning fails.

6.1 Introduction

Simultaneous localization and mapping (SLAM) is a fundamental problem in mobile robotics [29, 30]. Over the past several decades, efforts have been made to extend its application from laboratory scenes to outdoor scenes, where the major difference is the dynamic degree [2, 3, 32].

Self-positioning has been considered the hot spot in the research field of autonomous driving [4, 22], where existing methods can be divided into two groups according to the dependent information. The first type is to integrate the global information with internal sensors (e.g., odometer, IMU (inertial measurement unit), etc.) to dynamically adjust the self-pose estimation. However, the global information may become unstable and even lost due to the non-line of sight factor and the multi-path effect [9, 35], resulting in the cumulative deviation in dead reckoning (DR).

X. Zhang et al., *Multi-sensor Fusion for Autonomous Driving*,
https://doi.org/10.1007/978-981-99-3280-1_6

The second type of methods, simultaneous localization and mapping (SLAM), only requires local features (e.g., high-precision LiDAR in the Hector SLAM [20], odometry data in the Gmapping [12]) which provide higher accuracy and robustness. To overcome the cumulative deviation, several algorithms including the graph optimization strategy used in the Karto SLAM [21] and the optimal frame loop closure detection method in the Cartographer [18, 26] have been proposed. However, as SLAM-based methods depend on dense features in the surrounding environment, they fail to operate well in environments with sparse features, such as open areas or corridors.

In order to evaluate the performance of the proposed method, we constructed two simulation datasets with mobile robots in Sect. 6.2. At the same time, we also constructed two environments. The main feature in the first environment is the arc-shaped feature, and in the second environment, there are two corridors to compensate for each other's shortcomings with a hybrid positioning method. The robot used in our experiment has an encoder installed on each front wheel to provide displacement and angle increments. The IMU is located in the center of the robot to provide angular velocity; the single-line LiDAR is installed horizontally directly above the robot, with a horizontal scanning range of 360° and an angular resolution of 0.5°. In Sect. 6.3, we used a robot equipped with a 16-line Velodyne LiDAR, a camera, and an inertial measurement unit as a data collector to conduct data collection and verification in a parking lot. We guide the robot to run in the parking lot and collect data every few days to build a semi-dynamic map. To quantitatively compare the performance of this algorithm, the data evaluation criteria include average position error, maximum position error, position root mean square error, average angle error, and maximum angle error.

6.2 GF-SLAM

To leverage their advantage by integrating two types of methods, a straightforward strategy is to build up a high-definition map with SLAM, and the pose of the ego vehicle is then able to be updated according to the map when the global information is unstable or lost. For example, Choi et al. [8] detect lane lines, endpoints, and road signs to compare the high-precision map to correct the horizontal and vertical position, as well as the heading angle of the car. It can maintain high positioning accuracy even in the tunnel. Though the algorithm is reliable, it essentially requires collecting data and establishing a map in advance. Moreover, the maps limit the scope of action.

To avoid the limitations mentioned above, combining the global information with SLAM methods in real time is also considered [5, 7]. For example, Guivant [14, 15] applies GPS and LiDAR and updates the state vectors together with the feature points extracted from the environment. But the precision of the update process heavily depends on the stability of the overall system. Therefore, the lack of evaluation and screening of feature points may lead to unpredictable errors in the positioning. On the other hand, Song et al. [27] put multiple beacons into an extended Kalman filter (EKF) to update the coarse map and then build a fine local map by frame matching. In conclusion, these methods are based on global information and treat LiDAR as a local odometer to further correct the positioning pose. However, their implementation is complicated and requires high computing power, which constitutes the obstacle in the application.

In contrast to the existing methods mentioned above, we adopt the idea of coarse-and-fine hybrid information fusion for self-positioning and propose a system that integrates global information and feature-based SLAM (GF-SLAM). Neither stable global signals nor a pre-established map as the base of mapping [17] is required in the proposed method. When the system operates, the feature-based SLAM (F-SLAM) measures the features in the environment, including corner points, arcs, and line segments, to continuously construct the feature map for coarse positioning with a particle filter. During the process, each feature is filtered and updated independently to reduce the computation complexity. Moreover, if the global information is accessible, the system is converted to the precise positioning of EKF fusion DR. Simultaneously, the feature map is continuously updated on the results of EKF, which can gradually refine the coarse map of F-SLAM, thus eliminating the cumulative error in positioning. It should be noted that the global positioning would not affect the local F-SLAM directly, which ensures the stability of the system in a changeable environment.

To evaluate the performance of the proposed hybrid method [11], two virtual environmental scenarios are built, as well as the mobile robot used in Fig. 6.1. In the first environment, we aim to compare our F-SLAM with some state-of-the-art SLAM methods. Considering that the arc feature has better recognition and robustness, we set it as the main feature in this scene as shown in Fig. 6.2a. Meanwhile, we also build the second scene with two corridors to observe whether the hybrid positioning methods can compensate for each other's defects and switch stably under status with different global information. The results present that our F-SLAM can achieve a comparable performance of the mainstream SLAM algorithms and the integration of global information can further improve the mapping and positioning. The contributions of this chapter are listed as follows:

- The presented coarse-and-fine hybrid information fusion method is applied to the positioning systems and achieves robust performance under unreliable global signals.
- An improved feature-based SLAM is proposed, which approaches general performance with less computation than other SLAM systems.
- The positioning system is implemented to be real time and operates responsively to changes in the global signals, which can switch the algorithms accordingly.

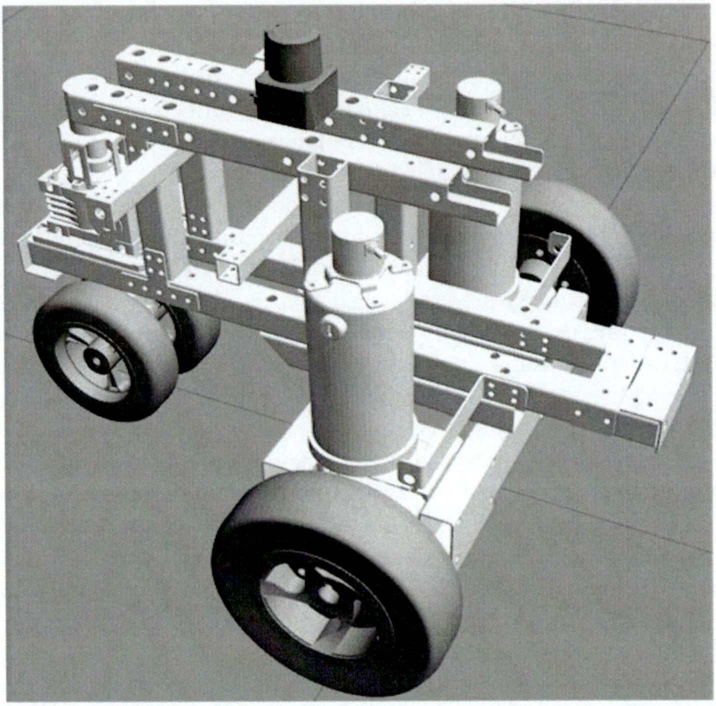

Fig. 6.1 The virtual model of the mobile robot, with two front-wheel drive and a single-rear-wheel steering structure, equipped with sensors such as single-line LiDAR, encoders, and IMU

6.2.1 Methodology

First, we present the process of fusing global information and motion models for fine positioning. It is converted to F-SLAM when the global information is lost. On the other hand, we introduce the establishment of the feature map and particle filter positioning (coarse positioning) in F-SLAM. The overview of our method is presented in Fig. 6.3.

The automated vehicle, which is represented by a mobile wheeled robot in our experiments, can build the corresponding motion models. In our setting, the robot runs at a low speed to simulate a two-wheel differential model. The global information and IMU data are the observations for EKF, constantly updating its pose for fine positioning.

We use the state vector $X_k = [x_k, y_k, \theta_k]^T$ to denote the robot's pose in the global coordinates at time k. The odometry motion model is:

$$X_k = f(X_{k-1}, w), w \sim N(0, Q), \tag{6.1}$$

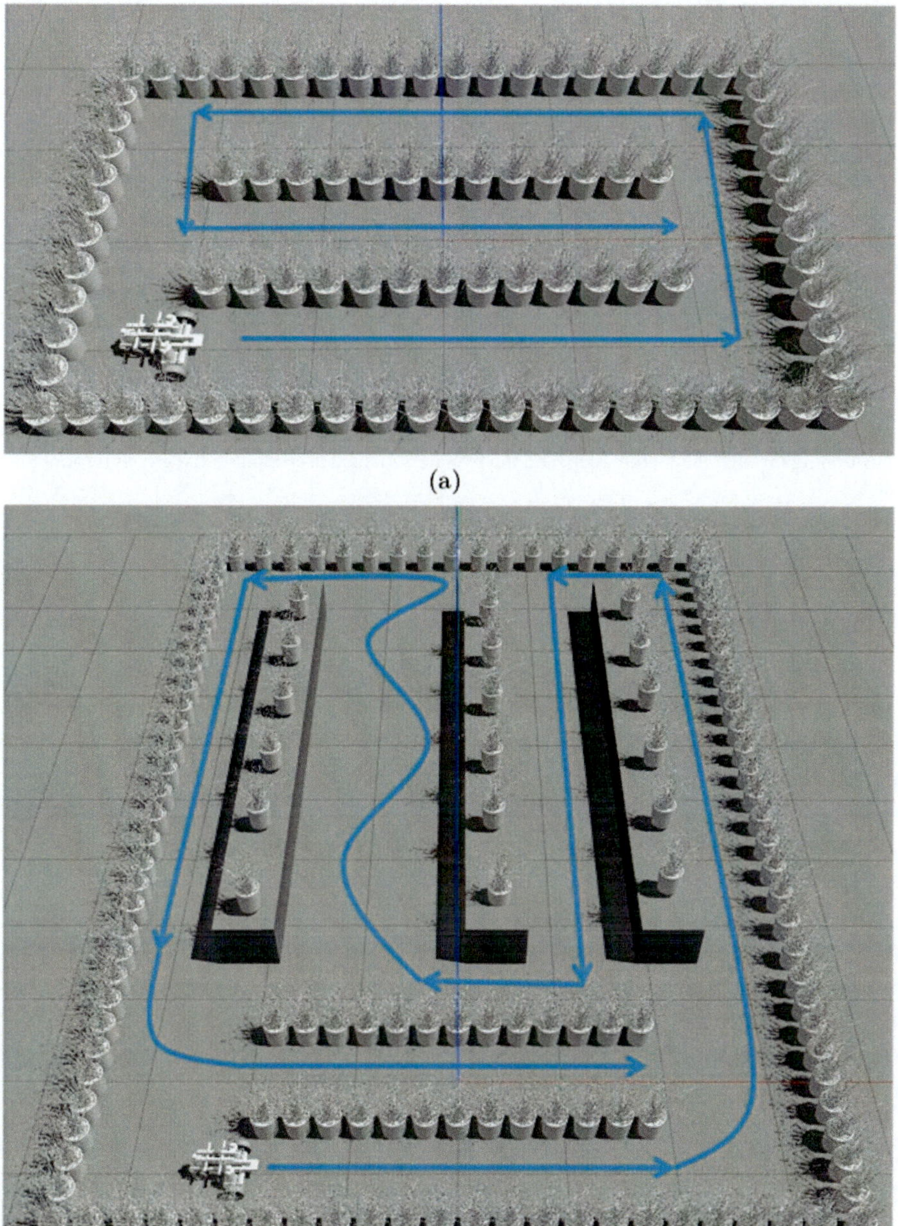

(a)

(b)

Fig. 6.2 The schematic sketch of two virtual environments. (**a**) Scenario I is dense with obstacles. (**b**) Scenario II with two corridors

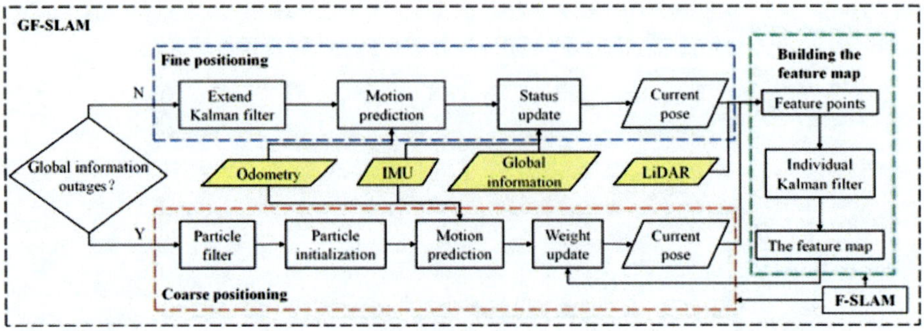

Fig. 6.3 Flowchart of the proposed method. The dark blue wireframe shows the EKF positioning using global information (fine positioning). The dark red wireframe and the dark green wireframe show the particle filter positioning (coarse positioning) and the flow of building the feature map, respectively, which together form F-SLAM. Fine positioning improves the accuracy of coarse positioning through the feature map, and the two are switched according to the global information state

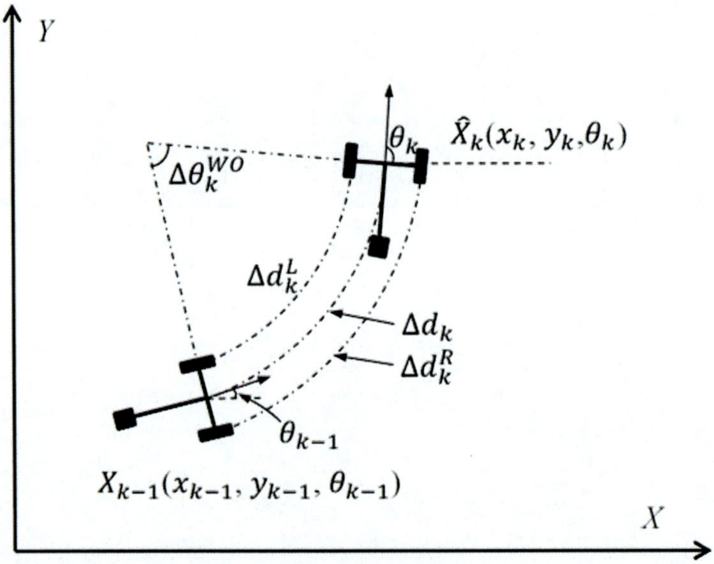

Fig. 6.4 The odometry motion model. θ_k is the positive angle with the X-axis at time k, representing the heading angle of the mobile robot

where w denotes the process noise with covariance Q, Δd_k, Δd_k^R and $\Delta\theta_k^{wo}$, as shown in Fig. 6.4 are defined as the displacement increment and the heading angle increment of the wheel odometer, respectively. They are defined as:

$$\Delta d_k = \frac{\delta d_k^R + \Delta d_k^L}{2}, \ \Delta\theta_k^{wo} = \frac{\delta d_k^R - \Delta d_k^L}{L}, \tag{6.2}$$

where δd_k^R and Δd_k^L denote the displacement increments of the left and right wheels within the sampling time, respectively. L denotes the distance between the wheels. Let $\hat{X}_k = [x_k, y_k, \theta_k]^T$ be the predicted state vector; then the transfer function can be defined as:

$$\begin{bmatrix} x_k \\ y_k \\ \theta_k \end{bmatrix} = \begin{bmatrix} x_{k-1} \\ y_{k-1} \\ \theta_{k-1} \end{bmatrix} + \begin{bmatrix} \frac{\Delta d_k}{\Delta \theta_k^{wo}}(\sin(\theta_{k-1} + \Delta\theta_k^{wo}) - \sin(\theta_{k-1})) \\ \frac{\Delta d_k}{\Delta \theta_k^{wo}}(\cos(\theta_{k-1} - \cos(\theta_{k-1} + \Delta\theta_k^{wo}))) \\ \Delta\theta_k^{wo} \end{bmatrix}, \tag{6.3}$$

The observation model is:

$$Z_k = h(X_k, v), v \sim N(0, R), \tag{6.4}$$

where $Z_k = [d_{1,k}, d_{2,k}, d_{3,k}, \theta_k]^T$ denotes the measurement vector and v denotes the measurement noise with covariance R. In our experiments, the positioning system is composed of three fixed ultrasonic transmitters with known positions $U_i = [x_i, y_i, z_i]^T, i = 1, 2, 3$. Let $d_{1\sim 3,k}$ be the distance between the robot and transmitters at time k, and the robot's global coordinates can be obtained. θ_k in Z_k is defined as follows:

$$\theta_k = \theta_{k-1} + \Delta\theta_k^{IMU}, \tag{6.5}$$

where the heading angle increment $\Delta\theta_k^{IMU}$ can be integrated within the IMU sampling time. The predicted measurement vector $\hat{Z}_k = [d_{1,k}, d_{2,k}, d_{3,k}, \theta_k]^T$ can be described as follows:

$$\hat{Z}_k = \begin{bmatrix} d_{1,k} \\ d_{2,k} \\ d_{3,k} \\ \theta_k \end{bmatrix} = \begin{bmatrix} \sqrt{(x_k - x_1)^2 + (y_k - y_1)^2 + (z_k - z_1)^2} \\ \sqrt{(x_k - x_2)^2 + (y_k - y_2)^2 + (z_k - z_2)^2} \\ \sqrt{(x_k - x_3)^2 + (y_k - y_3)^2 + (z_k - z_3)^2} \\ \theta_k \end{bmatrix}, \tag{6.6}$$

We assume that the moving surface is flat such that the height of the robot z_k is regarded as a constant. EKF time update equations:

$$\begin{aligned} \hat{X}_k &= f(X_{k-1}), \\ \hat{P}_k &= F_k P_{k-1} F_k^T + Q, \\ K_k &= \frac{\hat{P}_k H_k^T}{H_k \hat{P}_k H_k^T + R}, \\ X_k &= \hat{X}_k + K_k(Z_k - h(\hat{X}_k)), \\ P_k &= (I - K_k H_k)\hat{P}_k, \end{aligned} \tag{6.7}$$

where $\hat{P}_k$ denotes the prior state covariance matrix and P_k is a posteriori state covariance matrix. K_k and I are the Kalman gain and identity matrix, respectively. F_k and H_k are the Jacobian matrix of partial derivatives of the function f and H to x, respectively, and can be expressed as:

$$F_k = \frac{\partial f}{\partial X_{k-1}} + \begin{bmatrix} 1 & 0 & \frac{\Delta d_k}{\Delta \theta_k^{wo}}(\cos(\theta_{k-1} + \theta_k^{wo}) - \cos(\theta_{k-1})) \\ 0 & 1 & \frac{\Delta d_k}{\Delta \theta_k^{wo}}(\sin(\theta_{k-1} + \theta_k^{wo}) - \sin(\theta_{k-1})) \\ 0 & 0 & 1 \end{bmatrix}, \tag{6.8}$$

$$H_k = \frac{\partial h}{\partial X_k} + \begin{bmatrix} \frac{x_k - x_1}{D_{k,1}} & \frac{y_k - y_1}{D_{k,1}} & 0 \\ \frac{x_k - x_2}{D_{k,1}} & \frac{y_k - y_2}{D_{k,1}} & 0 \\ \frac{x_k - x_3}{D_{k,1}} & \frac{y_k - y_3}{D_{k,1}} & 0 \\ 0 & 0 & 1 \end{bmatrix}, \tag{6.9}$$

where $D_{k,i}$ is defined as:

$$D_{k,i} = \sqrt{(x_k - x_i)^2 + (y_k - y_i)^2 + (z_k - z_i)^2}, i = 1, 2, 3 \tag{6.10}$$

During the move, the mobile robot performs preprocessing and feature extraction for each laser scan, as well as updating each feature point in the global coordinates. To preserve good historical observations, our algorithm filters and updates each feature individually while eliminating those with poor historical observation quality. In this way, the generated feature map only occupies little memory. When global information deteriorates, the system is then converted to particle filter positioning under the feature map to update the pose. It initialized the particle swarm with the EKF estimation at the previous moment and combined it with the F-SLAM, which does not require global information but is simple and efficient.

The proposed method can adaptively switch between two positioning modes according to the global information state, maintaining the coherence and stability of the overall positioning path. The two positioning methods interact through the feature map. Though F-SLAM maintains accurately in a short time, as a coarse positioning method, its minor deviation accumulates and gradually becomes unignorable. Once global information is available and converted to fine positioning, both map and position errors will be corrected. We will detail the process in the following.

Extraction and transformation of feature points. After preprocessing the single frame scan, which may contain features such as arcs and corner points, we obtain local coordinates of the feature points $[x_{L,k}^i, y_{L,k}^i]^T, i = 1, \ldots, N_F$ (where N_F represents the number of feature points) using circle fitting, segmented line fitting, etc. Based on the current state of the global information state, the pose of our robot can be obtained by EKF or particle filter, as well as the pose of the LiDAR which can be obtained from the installation position relationship. Local coordinates of feature points are then converted to the global coordinate system $[x^i_{G,k}, y^i_{G,k}]$ by using the LiDAR position as a base point. When LiDAR is installed in the horizontal center position, and the initial scanning beam is directly in front of the LiDAR, the coordinate transformation of feature points can be defined as:

$$\begin{bmatrix} x^i_{G,k} \; y^i_{G,k} \end{bmatrix} = [x_k \; y_k] + \begin{bmatrix} \cos_{\theta_k} & -\sin_{\theta_k} \\ \sin_{\theta_k} & \cos_{\theta_k} \end{bmatrix} \begin{bmatrix} x^i_{L,k} \\ y^i_{L,k} \end{bmatrix} \tag{6.11}$$

When all the feature points are converted, they are collected in the current feature point set C_k.

Registration and updating of features. EKF-SLAM adds all the extracted features to the state vector and updates it with the vehicle state. The design and the error estimation matrix are complicated [8, 28]. Therefore, we build an individual Kalman filter for each feature. We define the recorded feature point set S_k, which contains multiple features S_i. The global coordinate of S_i is its state vector $B^i_k = [x^i_k, y^i_k]^T$ (where the subscript k represents time slot and i denotes the feature number). Initialize S_k with the C_k obtained from the scan of the first frame, i.e., add all the feature point state vectors to S_k. When the new C_k arrives, traverse S_k for nearest-neighbor matching. Only those points with a smaller distance than the matching threshold d_match will be considered as successful matchings. Failed feature points will become new features, while successfully matched feature points are used as the observation to update the corresponding feature state vectors. The motion and observation models are set individually as below:

$$\begin{aligned} B^i_k &= F_B B^i_{k-1} + w_B, w_B \sim N(0, Q_B), \\ E^i_k &= H_B B^i_k + v_B, v_B \sim N(0, R_B), \end{aligned} \tag{6.12}$$

where E^i_k denotes new observations for a feature. Since features are static, the expressions for the observations and state vectors are equal, and the state-transferring matrix and observation matrix H_B are both unit matrices. The observations of features are influenced by both LiDAR and positioning errors, and the process noise matrix Q_B and the measurement noise matrix R_B can be adjusted by the results of multiple experiments. As a result, whenever a feature state vector receives a new matching feature point, the Kalman filter can be updated. The modification simplifies the complex error estimation matrix in filtering and avoids poor observation of features, which further contributes to the positioning.

Evaluation and screening of features. In the EKF-SLAM, a single poorly observed feature can affect the overall state matrix, and the lack of feature discrimination will cause unnecessary computational burden and errors. To evaluate the reliability of feature vectors, we apply N^i_A to denote the total number of times within the detection range of LiDAR and N^i_v to denote the number of times S_i

Algorithm 1 Update the feature map

if S_k does not exist, initialize S_k with C_k obtained from the scan of the first frame. **then**
    while $C_k \neq \Phi$ **do**
        Pop c from C_k.
        for s_i in S_k **do**
            Nearest-neighbor matching.
        end for
        if minimum distance between s_i and $c < d_{match}$ **then**
            Update s_i with c by individual Kalman filter.
            if $N_v^i > N_{\min}$ **then**
                Add c to S_k
            end if
        else
            Add c to T_k.
        end if
    end while
    for s_i in S_k **do**
        if $N_V^i < 20$ and $r_V^i < 0.5$ **then**
            Delete s_i
        end if
    end for
end if
return S_k, T_k

successfully matches a feature point. A high observation quality can be guaranteed only after accumulating several effective observations of a feature point, and we set $r_V^i = N_V^i / N_A^i * 100\%$ to avoid collecting dynamic obstacles in the environment as features. Removing a feature when its N_V^i and r_V^i are less than the preset threshold can avoid infinite expansion of S_k. When $N_V^i > N_{\min}$ ($N_{\min}$ is the minimum number of valid observations), the feature is considered to be stable and reliable and can be used for matching in the subsequent particle filter. All eligible features constitute the feature map. The establishment and updating of the feature map T_k are shown in Algorithm 1.

However, positioning and mapping in F-SLAM influence each other. The lack of loop detection may produce cumulative errors after long-term operation. Therefore, mapping is carried out during the fine positioning as well, and the high-precision base point can greatly improve the map quality, thereby eliminating the mapping error during the coarse positioning. Figure 6.5 visualizes the feature map in the experimental Scenario I and displays it together with maps created by other methods.

Judging global information outages. In the ultrasonic positioning systems, when the signal is unstable or obscured, the distance measurement will jump, causing a large error in the EKF optimization pose, while the position derived from the odometry model is continuous during the travel of the mobile robot. We define ΔD_k to denote the Euclidean distance between these two positions at time k and $\Delta\alpha_k$ to denote the angle between the connection of the two positions and the attitude of the odometry model. When ΔD_k >0.05m or $\Delta\alpha_k$ >10°, it indicates that the global

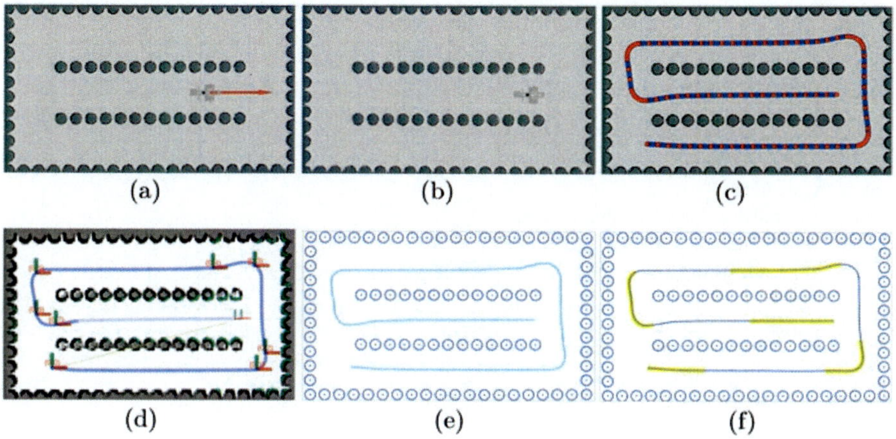

Fig. 6.5 Mapping display of different methods in Scenario I. (**a**) Hector SLAM. (**b**) Gmapping. (**c**) Karto SLAM. (**d**) Cartographer. (**e**) F-SLAM. (**f**) GF-SLAM. Yellow sections indicate global information outages, and blue circles indicate fitted arc features, and same meaning in Figs. 6.7, 6.8, and 6.9

information is unattainable, and the system switches to particle filter for positioning, provided it is available. Otherwise, if particle filter is unavailable, the system outputs the pose directly based on the odometry model.

Initialization of the particle swarm. The particle swarm is initialized with X_{k-1} as the center and P_{k-1} as the proposal distribution for sampling. By randomly picking M initial particles, $X_k^i = [x_k^i, y_k^i, \theta_k^i]$ denotes the state of the ith particle with the weight $w_k^i = 1/M$.

Particle prediction and weight updating. We apply the same motion model as EKF positioning for the prediction of each particle. Compared with the encoder, IMU calculates the angle increment more accurately, so $\Delta\theta_k^{w_0}$ is replaced by $\Delta\theta_k^{IMU}$ directly. Feature matching of particle swarms takes up significant runtime for weight update. As an improvement, we first use the weighted average of the particle swarm after motion prediction to obtain a matching particle X_k^m and then compare X_k^m with the feature map to get matchable features; if the number of matchable features is less than 3, the position may have a large error or even jump. In this case, X_k^m is directly outputted as the position. If the number of matching features is sufficient, the observation error of each particle is calculated to update its weight. The weight update is calculated according to the following definition:

$$w_k^i = w_{k-1}^i \prod_{n=1}^{N} \frac{1}{2\pi\sigma_x\sigma_y} \exp^{-\frac{1}{2}((\frac{\Delta\overline{x}}{\sigma_x})^2+(\frac{\Delta\overline{y}}{\sigma_y})^2)} \tag{6.13}$$

Algorithm 2 Particle filter positioning under feature map

while S_k does not exist, initialize S_k with C_k obtained from the scan of the first frame. **do**
  Initialize the particle swarm with X_{k-1} and P_{k-1}
  for i=1 to M **do**
    Compute the motion model for particle X_k^i.
  end for
  Calculate X_k^m
  Get the current matchable features.
  if the number of matching features <3 **then**
    Output X_k^m.
  else
    for $i = 1$ to M **do**
      Compute the observation error of particle i.
      Update the weight.
    end for
    Weight normalization.
    if N_{eff} <$2M/3$ **then**
      Low-variance resampling.
    end if
    Calculate X_k,ΔD_k and $\Delta\alpha_k$ between X_k and X_k^m.
    if ΔD_k >0.05m or $\Delta\alpha_k$ >10° **then**
      Output X_k^m.
    else
      Output X_k and calculate P_k.
    end if
  end if
end while

where w_k^i denotes the weight of the th particle, N is the number of current matchable features, and $\Delta\overline{x}$ and $\Delta\overline{y}$ are the observation error of the ith particle to the feature. After the weights are updated, they are normalized to sum up to one.

Resampling and output. Whether to resample is based on the number of valid particles N_{eff}, which can be given by:

$$N_{eff} = \frac{1}{\sum_{i=1}^{M}\left(\widetilde{w}_k^i\right)^2} \tag{6.14}$$

where $\widetilde{w}_k^i$ represents the normalized weight. If $N_{eff} < \frac{2}{3}M$, we perform low-variance resampling and then output the pose as:

$$X_k = \Sigma_{i=1}^{M}\tilde{w}_k^i X_k^i \tag{6.15}$$

Because particle filter is frequently used in situations where discontinuous positioning occurs, the method for assessing global information outages is applied to calculate ΔD_k and $\Delta\alpha_k$ between the latest position estimate X_k and the previous position estimate X_k^m. If the difference is deemed significant, this indicates a "jump" has occurred, which prompts the output of X_k^m; otherwise, X_k is output. Calculate

P_k with X_k as the center and particle swarm as the distribution, to be ready to switch to EKF positioning. Algorithm 1 gives the pseudo-code of particle filter positioning.

6.2.2 *Experiment*

The mobile robot used in our experiment is shown in Fig. 6.1. Encoders are installed on each of the front wheels to provide displacement and angular increments, and an IMU is located in the center of the robot to provide angular velocity, both of which have an error of 1% per frequency. A single-line LiDAR is horizontally installed directly above the robot, the horizontal scanning range is 360°, and the angular resolution is 0.5°. To improve the efficiency and quality of feature extraction, only the points within 5m from the scanning center are retained, with a distance error of 0.5%. In the environment, three ultrasonic transmitters are installed to provide global information. Since we only focus on the two-dimensional pose, the ultrasonic transmitters are set to the same height. They all have a distance Gaussian noise error of 2 cm, and the average position error of the solution is about 3 cm. To facilitate the research, the signal frequencies of all the above sensors are set to 10 Hz.

As the arc features are easy to identify and fit, also with good robustness, we only set the arc features in the experimental scenarios. We use flowerpots as obstacles. When the obstacles are densely distributed, the ultrasonic signal will be lost due to the occlusion of branches and leaves, thereby simulating global information interruption. Other methods are also run in this scenario to verify the effectiveness of the proposed method.

We implemented the experiment with the ROS of version Melodic. The mobile robot goes in the planned route and records raw data from each sensor into the rosbag. In the experiment, the mapping and positioning were compared with several typical SLAM methods, Hector SLAM, Gmapping, Karto SLAM, and Cartographer, all of which can be directly configured in ROS. We read the rosbag in MATLAB 2019b and compiled the entire algorithm program. The positioning results of other theories are also quantified, compared, and visualized in MATLAB.

To quantitatively compare various algorithms, we use the mean position error $\overline{e}_d$, maximum position error e_{dmax}, root mean squared error of position RMSE, mean angular error $\overline{e}_a$, and maximum angular error e_{amax} as evaluation metrics for positioning, taking the average results of various methods performed 20 times. The algorithm parameters in the experiment are as follows: $L = 0.46$m, d_{match} can't be greater than the minimum spacing of adjacent features in the environment, and the minimum distance between obstacles in the scenario is 0.2m; thus, we set $d_{match} = 0.1$m and $N_{\min} = 20$. The number of particles M is set to 250. The process noise and measurement noise used in EKF positioning and feature filtering have been experimentally adjusted to the following values:

$$Q = \begin{bmatrix} (0.001)^2 & 0 & 0 \\ 0 & (0.001)^2 & 0 \\ 0 & 0 & 0.1 \end{bmatrix},$$

$$R = \begin{bmatrix} (0.02)^2 & 0 & 0 & 0 \\ 0 & (0.02)^2 & 0 & 0 \\ 0 & 0 & (0.02)^2 & 0 \\ 0 & 0 & 0 & (0.02)^2 \end{bmatrix}, \tag{6.16}$$

$$Q_B = \begin{bmatrix} (0.0001)^2 & 0 \\ 0 & (0.0001)^2 \end{bmatrix},$$

$$R_B = \begin{bmatrix} (0.02)^2 & 0 \\ 0 & (0.02)^2 \end{bmatrix},$$

Scenario I has a regular distribution of obstacles and rich features, which is very suitable for the operation of various SLAM methods and testing the effectiveness of F-SLAM. The robot moves along the trajectory indicated by the blue arrow in Fig. 6.2, with the maximum linear velocity under 1m/s, and the overall process is about 110 s. Though the F-SLAM is simple to implement, the positioning effect can be easily affected by the initial position. Because the subsequent feature map updating and positioning are obtained based on the initial observation, the initial observation quality must be guaranteed. After the simulation starts, the robot is stationary at the start point for 2–3 s to ensure that the initial feature position can be recorded stably. Then, the robot starts to move along the planned route.

GF-SLAM is also tested in this scenario by randomly setting a few sections of global information outages, which accounts for about 48% of the total path. Using the same rosbag, each algorithm runs independently 20 times; the parameters of several algorithms are well adjusted to ensure the stability of the results. We also include the EKF+DR method in the comparison, that is, the EKF positioning is performed only with stable global information, and the positioning can be supplemented by DR during global information outages. We select the positioning results with the moderate performance of each algorithm. The comparison of positioning paths and cumulative error is performed in Figs. 6.6 and 6.7, respectively, and the results of evaluation metrics for each SLAM method are recorded in Table 6.1.

From mapping and path comparison, various SLAM methods can build complete maps in feature-rich scenes, and the offset between the positioning path and ground truth is kept within 2cm. The position and heading angle accuracy of Gmapping are the highest among several methods, and Hector SLAM can achieve sub-optimal positioning results without odometry, which indicates that the SLAM methods perform well under rich environmental features. The values of $\bar{e}_d$, RMSE, and $\bar{e}_a$ of F-SLAM are closed to Hector SLAM, but are smaller than Karto SLAM with a

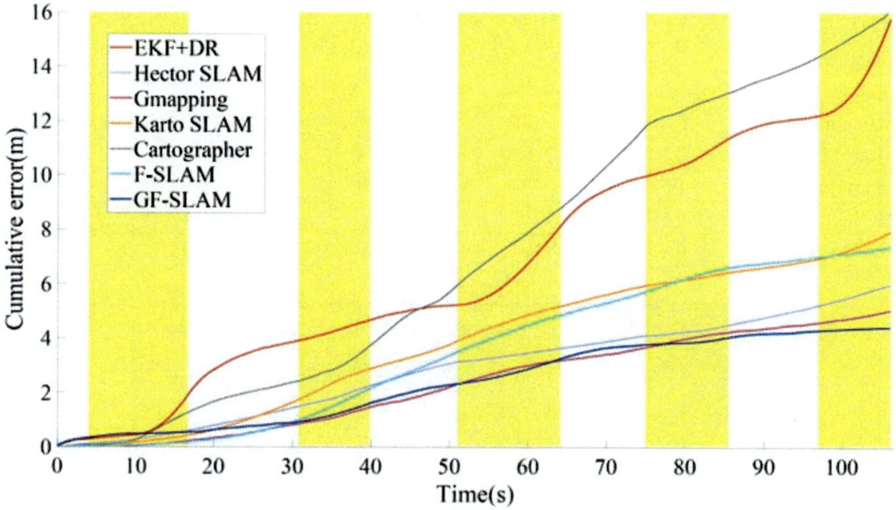

Fig. 6.6 Cumulative error comparison in Scenario I. The yellow sections indicate global information outages, and same meaning in Figs. 6.10 and 6.11

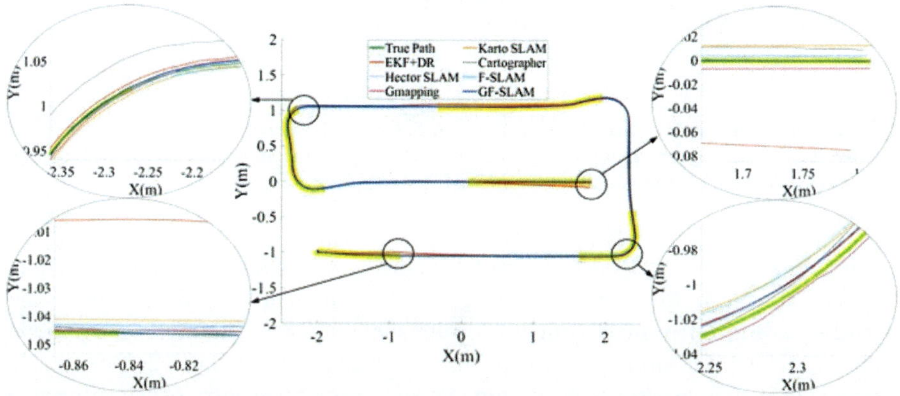

Fig. 6.7 Path comparison in Scenario I. According to the lines, the proposed F-SLAM can achieve similar precision as other SLAM methods. Besides, GF-SLAM performs better with the correction through global information when it is available

flatter growth of the whole process cumulative error. The results show that F-SLAM can successfully be mapping and achieving similar positioning accuracy to other SLAM methods under rich features. In addition, GF-SLAM has performed a total of nine positioning switches, whose $\bar{e}_d$ and RMSE are further reduced, indicating that global information can improve position accuracy, but its e_d and e_{amax} are larger than other methods, because other SLAM methods are more sensitive to changes in angle due to frame matching and feature registration.

Table 6.1 Quantitative comparison of methods

Methods	$\overline{e}_d$(cm)	e_{dmax}(cm)	$RMSE$(cm)	$\overline{e}_a$(°)	e_{amax}(°)
Hector SLAM	0.59	1.24	0.63	0.14	0.69
Gmapping	0.49	2.14	0.6	0.2	0.58
Karto SLAM	0.94	2.3	1.06	0.4	1.2
Cartographer	1.52	3.71	1.76	0.24	1.22
F-SLAM	0.69	1.75	0.81	0.17	0.79
GF-SLAM	0.43	1.4	0.54	0.49	3.11

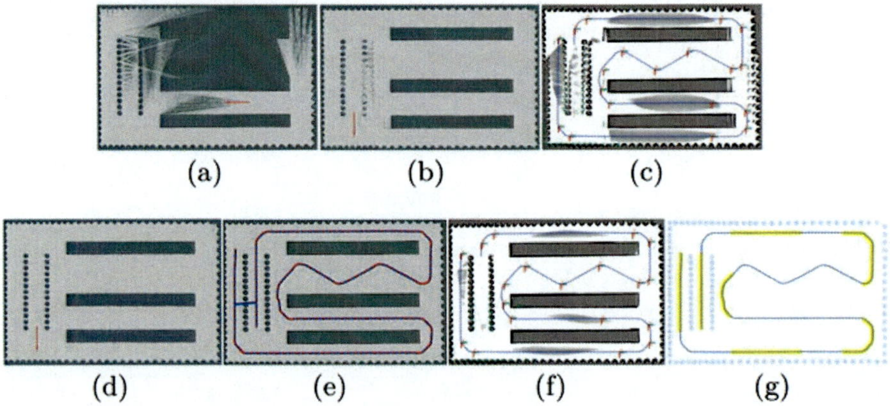

Fig. 6.8 Mapping display of different methods in Scenario II. (**a**) Hector SLAM mapping drift and overlap. (**b**) Gmapping mapping overlap. (**c**) Cartographer mapping overlap. (**d**) Gmapping. (**e**) Karto SLAM. (**f**) Cartographer. (**g**) GF-SLAM

In Scenario II, we added two 7-m-long corridors in which the lack of distinct features may lead to poor performance of conventional SLAM. However, the corridors are in open areas with stable ultrasonic signals. Therefore, our mobile robot executes F-SLAM for coarse positioning when it is located in a dense obstacle area and then switches to fine positioning when traveling to a corridor. The objective is to check the stability of the two positioning methods when switching and the effect on F-SLAM after introducing global information. The robot moves along the indicated trajectory for about 270 s. The trajectories in the two long corridors are straight or curve like an "S." We only provide ultrasonic signals in corridors and areas with sparse obstacle distribution, and the rest of the scenario is set to global information outages, which accounts for about 54.7% of the total path. To verify whether the lack of features affects other SLAM algorithms, we attempt to reduce the detection range of LiDAR to 3m. Thus, while moving through the corridor, the system can only scan two parallel lines within a certain period. Figure 6.8a–c shows several cases of mapping or positioning failure. After adjusting the LiDAR detection range to 5m, when the robot turns sharply or stays in the corridor, Hector SLAM will still produce the map to drift or stop positioning, while other methods can successfully perform mapping and positioning. Their maps are presented in

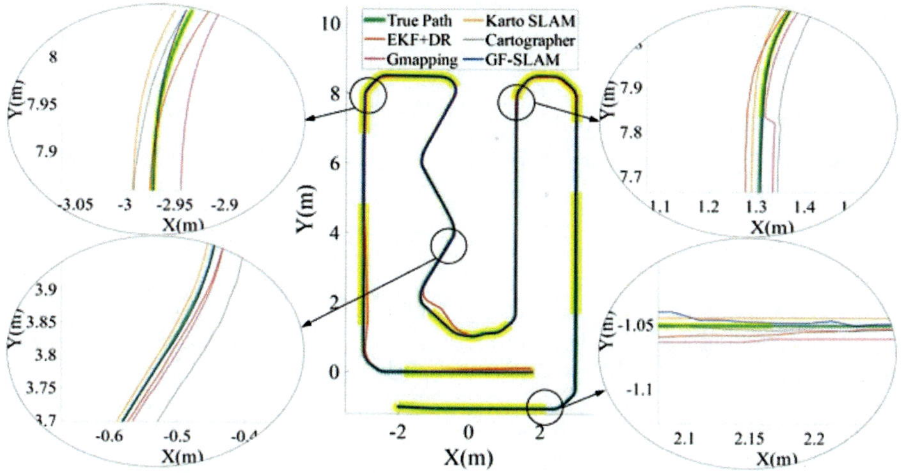

Fig. 6.9 Path comparison in Scenario II. The path is enlarged and compared in several coarse-and-fine positioning switching areas, and it can be seen that the path of GF-SLAM is smooth and stable

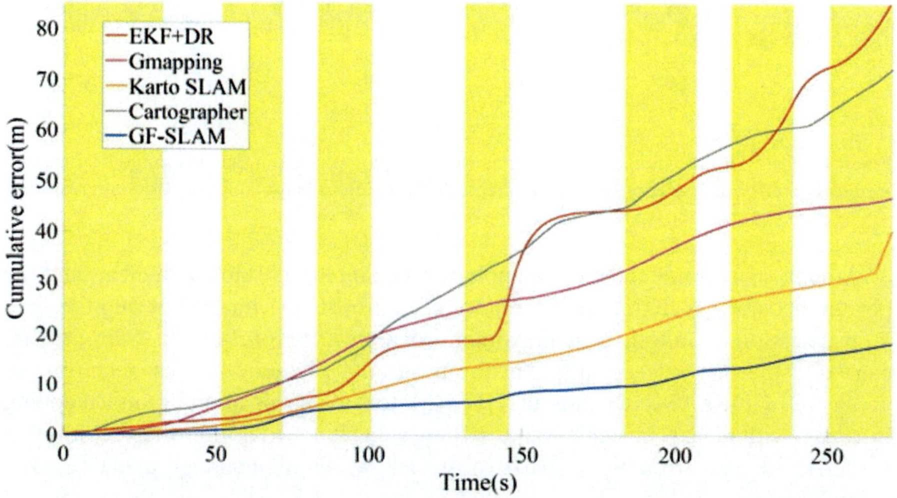

Fig. 6.10 Cumulative error comparison in Scenario II. The cumulative error of EKF+DR increases rapidly after the interruption of global information, and the change of GF-SLAM is slow, with the smallest cumulative error

Fig. 6.8d–g, where the path and cumulative error comparisons are performed in Figs. 6.9 and 6.10, respectively. The evaluation metrics of methods that can be completely mapped are recorded in Table 6.2. Figure 6.11 plots the variation in position error for GF-SLAM.

Table 6.2 Quantitative comparison of methods

Methods	$\overline{e}_d$(cm)	e_{dmax}(cm)	$RMSE$(cm)	$\overline{e}_a$(°)	e_{amax}(°)
Gmapping	1.7	4.1	1.92	0.17	3.16
Karto SLAM	1.57	14.8	2.64	0.13	4.04
Cartographer	2.7	6.75	3.17	0.3	1.5
GF-SLAM	0.79	3.31	1.19	0.51	3.73

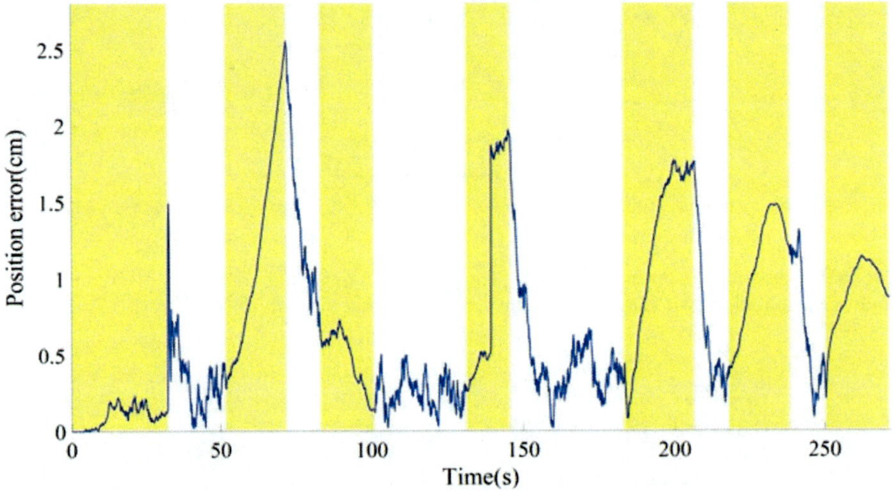

Fig. 6.11 GF-SLAM position error in Scenario II. No jump occurs at the positioning switch, and the position error of coarse positioning decreases rapidly after the global information regression

Among several methods, the mapping of Gmapping is stable and clear, and the posture accuracy is high, while the pose of Karto SLAM jumped during the final loop detection, resulting in a maximum position error of 14.8cm. Cartographer can build the map successfully, but the positional accuracy is poor according to Table 6.2. GF-SLAM switches to F-SLAM for coarse positioning and mapping in areas with dense obstacles and switches back to EKF for fine positioning in corridors and areas with sparse obstacles while maintaining mapping. The enlarged image in Fig. 6.9 shows that except for EKF+DR, the trajectories of other methods are kept near the ground truth most of the time. From Fig. 6.10, the cumulative error of EKF+DR increases rapidly after global information is interrupted, and the growth curve of GF-SLAM is smoother than other methods. $\bar{e}_d$, e_{dmax}, and RMSE of GF-SLAM are smaller according to Table 6.2, which indicates that our method can effectively utilize the global information to improve the positioning accuracy. Figure 6.11 indicates that the whole process occurred 12 times of positioning switching and no jumps occurred before and after the switchover, achieving seamless switching of coarse and fine positioning. Due to lack of correction, there are cumulative errors after F-SLAM runs for a long time. When the global information returns, the position error is rapidly reduced, and the cumulative error is eliminated. Both $\bar{e}_a$ and e_{dmax} of GF-SLAM are higher than other SLAM methods. The reason is that the accuracy of using EKF to estimate the heading angle is not too high. But the overall heading angle error is still within an acceptable range.

6.3 Lifelong Localization in Semi-Dynamic Environment

Mapping and localization in non-static environments are fundamental problems in robotics. Most previous methods mainly focus on static and highly dynamic objects in the environment, which may suffer from localization failure in semi-dynamic scenarios without considering objects with lower dynamics, such as parked cars and stopped pedestrians. In this section, semantic mapping and lifelong localization approaches have been proposed to recognize semi-dynamic objects in non-static environments. We also propose a generic framework that can integrate mainstream object detection algorithms with mapping and localization algorithms. The mapping method combines an object detection algorithm and a SLAM algorithm to detect semi-dynamic objects and constructs a semantic map that only contains semi-dynamic objects in the environment. During navigation, the localization method can classify observation corresponding to static and non-static objects, respectively. Then our method evaluates whether those semi-dynamic objects have moved, to reduce the weight of invalid observation and localization fluctuation. Real-world experiments show that the proposed method can improve the localization accuracy of mobile robots in non-static scenarios.

Besides static objects, more frequently seen dynamic and semi-dynamic objects in outdoor environments may cause fluctuations in the localization system, which pose crucial challenges to mapping and localization system [1, 10, 16, 23]. However, in environments containing many semi-dynamic objects, mapping and localization are still open problems and remain unsolved due to the similarity of dynamic characteristics of static and semi-dynamic objects. Therefore, in this section, we propose a related method to recognize and record the static and semi-dynamic objects in the environment, which can be divided into two parts: mapping static objects and semi-dynamic objects, respectively, and robot localization using both static maps and semi-dynamic maps (Fig. 6.12).

The mapping algorithm in this work extends the occupancy grid technique introduced by Hess et al. [19] to record semi-dynamic parts of the environment. It is capable of detecting both static and semi-dynamic objects in the environment and generating two maps: the static map and the semi-dynamic map. After filtering out the dynamic objects during the mapping process, the static map contains both static and semi-dynamic objects in the environment, while the semi-dynamic map only contains semi-dynamic objects.

The localization algorithm in this work optimizes the particle filter algorithm by using two maps simultaneously. It allows for the observation classification (static or semi-dynamic objects), as well as a determination of the distance the semi-dynamic

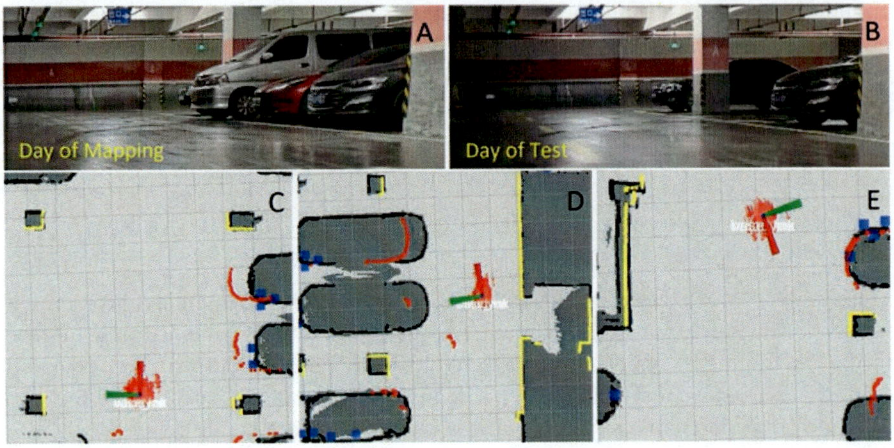

Fig. 6.12 Robot operates in the environment with both static objects (walls and pillars) and semi-dynamic objects (cars). The environments during mapping (**a**) and localization (**b**) are quite different due to the moving of cars. The figures (**c**, **d**, **e**) represent the observation during the localization process; the red LiDAR points correspond to semi-dynamic objects, which may or may not have been moved. The yellow LiDAR points correspond to static objects. The weight of the current pose can be recalculated by the classification of observation

objects have moved. Followed by adjustment of the weights of different objects, a more precise localization result can be achieved in semi-dynamic environments.

Compared with traditional features, semantic features are invariant against environmental changes. Vineet et al. [31] introduced scene understanding to large-scale semantic scene reconstruction. They also presented a semantic fusion approach that can handle dynamic objects more effectively. Yu et al. [36] combined semantic segmentation network with moving consistency to improve the localization accuracy. Wang et al. [33] projected LiDAR data into images and applied 2D segmentation information of RGB images to 3D LiDAR points. Chen et al. [6] proposed a pose optimization method based on semantic features that simultaneously adjusted the tree position while estimating the robot pose. Previous semantic information-based methods have been successfully applied in dynamic scenes, but neglecting the semi-dynamic characteristics represents the major limitation. In this chapter, we propose a system that contains a mapping approach that utilizes an object detection algorithm to generate a semantic map. The following localization algorithm can be applied in semi-dynamic environments.

In summary, our main contributions are:

- A semantic mapping method was proposed that can identify and record the environment's semi-dynamic features.
- We construct a robust localization system that can determine object categories according to the observation.

- We also introduce a generic framework that can combine mainstream object detection and SLAM algorithms to achieve high localization accuracy regardless of position changes of semi-dynamic objects.

6.3.1 Methodology

This section describes the architecture of the proposed method. According to the classification from Meyer-Delius et al. [24], in this section, the classifications are slightly revised as follows:

- Static objects: objects that never change their position, like walls and shelves
- Semi-dynamic objects: objects that are static during the mapping process but may be moved during the localization process, like chairs and parked cars
- Dynamic objects: objects that change their position frequently, like moving people and moving cars

Figure 6.13 gives an overview of the mapping and localization framework. During the mapping process (red dashed line), an object detection algorithm (Yolov3 [25]) is selected to recognize the semi-dynamic objects. In the SLAM thread, we exploit a SLAM algorithm (Cartographer [19]) to generate a static map, followed by projecting LiDAR points into the image. The given semi-dynamic position

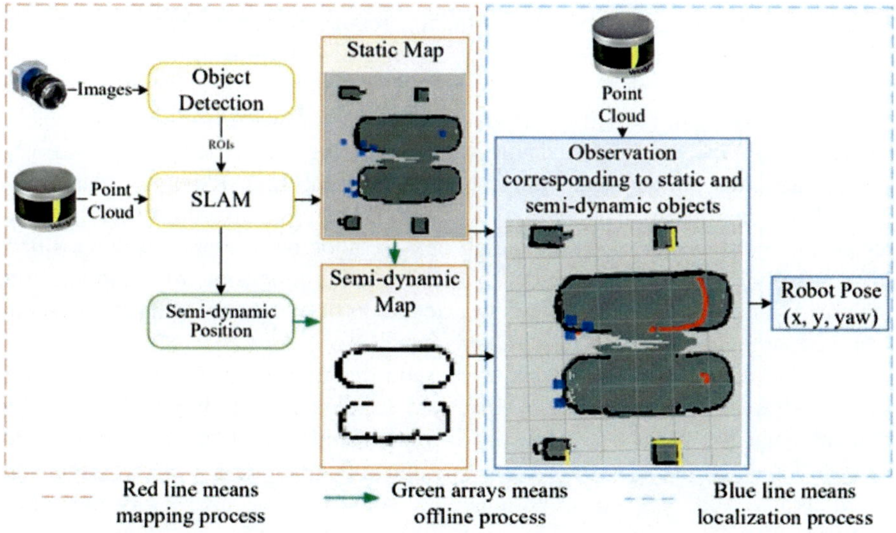

Fig. 6.13 The architecture of the proposed mapping and localization system. The static map contains two cars and four pillars, and the semi-dynamic map contains two cars. The blue dots in maps represent the semi-dynamic positions. The red LiDAR points and yellow LiDAR points correspond to semi-dynamic objects and static objects, respectively

in global coordinate can later facilitate creating a semi-dynamic map m_d from the static map m_s. The static map m_s contains information of both static and semi-dynamic objects, while the semi-dynamic map m_d only contains information of semi-dynamic objects. Moreover, the semi-dynamic map is generated offline without increasing the complexity of the mapping process. During the localization process (blue dashed line), m_s and m_d are simultaneously loaded to the localization system. By comparing the distance of observation, we can obtain observations corresponding to static objects (yellow points) and semi-dynamic objects (red points). Information from both m_s and m_d is incorporated to localize the robot's pose, depending on whether and how far the object is moved. Precise localization can be achieved by reducing the weight of observation corresponding to semi-dynamic objects.

The recognition of dynamic objects is achieved based on the probability that objects appear at the same location. Specifically, a certain area is dynamic if its status changes between occupied and free. In this chapter, a probabilistic method was utilized to define the problem as:

$$P(s_t|z_{1:t}, s_{t-1}) \tag{6.17}$$

where $z_{1:t}$ is the sensor data from time 1 to time t, s_{t-1} is the previous static map, and s_t is the current static map. By applying Bayes rule to Eq. (6.17) and eliminating some unnecessary terms, we can get an inverse observation model that can only record the static parts of environments. Specifically, the grid will be recorded if the status of a grid is changing from unknown to occupied or is staying occupied. Please refer to [19, 34] for more detailed information about inverse observation model.

Observation z_t is classified into observation of static objects z_t^s or observation of semi-dynamic objects z_t^d. For this step, Yolov3 was applied to get the classification information. By calibrating the extrinsic parameters between LiDAR and camera, LiDAR points can be projected into these bounding boxes. However, it cannot provide accurate dimensional information because the LiDAR points at the edge of the bounding box may correspond to other objects. To solve this problem, only the points in the center of the bounding box are adopted to represent the position information of semi-dynamic objects, which excludes dimension information. With the position information, the following calculation of connected areas on the static map gives a more accurate semi-dynamic map.

The blue dots in Fig. 6.14 indicate the semi-dynamic positions. One or several connected areas indicate a single object. To supplement the dimension of semi-dynamic objects, we use a relatively low threshold to obtain the binary static map, followed by calculation of connected area. The integration of semi-dynamic positions and connected areas produces the semi-dynamic map. Only the semi-dynamic positions are obtained when constructing the static map; other processes in Fig. 6.14 are offline which will not increase the computation complexity in the mapping process.

For the localization process, the Monte Carlo method is used to validate the proposed idea by using odometry u_t and observation z_t recursively. We improve the

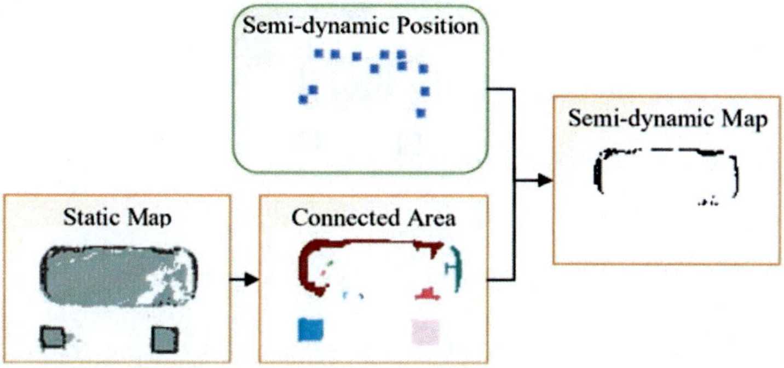

Fig. 6.14 A schematic diagram of the proposed mapping system. Semi-dynamic position (blue dots) and connected area (colored static map) produce the semi-dynamic map

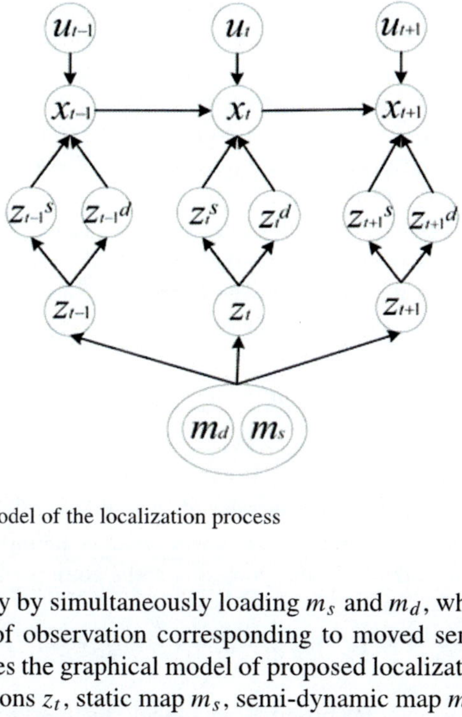

Fig. 6.15 Graphical model of the localization process

localization accuracy by simultaneously loading m_s and m_d, which can identify and reduce the weight of observation corresponding to moved semi-dynamic objects. Figure 6.15 illustrates the graphical model of proposed localization approach, which has sensor observations z_t, static map m_s, semi-dynamic map m_d, odometry u_t, and pose at time $t-1$ as input. The localization problem can be denoted as:

$$P(x_t|z_t, u_t, x_{t-1}, m_s, m_d) \tag{6.18}$$

By applying Bayes rule to Eq. (6.18), it can be expressed as:

$$\begin{aligned} &P(x_t|z_t, u_t, x_{t-1}, m_s, m_d) \\ &= \eta P(z_t|x_t, m_s, m_d) P(x_t|u_t, x_{t-1}) \end{aligned} \tag{6.19}$$

where η is a normalization constant. To further reduce the weight of moved objects, the classification of observation should be calculated based on m_s and m_d. Equation (6.19) can further be defined as:

$$\begin{aligned} & P(x_t|z_t, u_t, x_{t-1}, m_s, m_d) \\ & = \eta P(z_t^s, z_t^d|x_t, m_s, m_d)P(x_t|u_t, x_{t-1}) \end{aligned} \tag{6.20}$$

The major improvement of the proposed method is that two maps are used simultaneously to classify the observation corresponding to semi-dynamic objects and static objects. The classification can be calculated by comparing the nearest distance between each LiDAR point and two maps. Let d_i^s represent the distance from point i to the nearest grid in the static map, and let d_i^d represent the distance from point i to the nearest grid in the semi-dynamic map. If the difference between d_i^d and d_i^s is smaller than a threshold, current LiDAR points correspond to a semi-dynamic object; otherwise, the LiDAR point corresponds to a static object. The given result during the localization process (Fig. 6.13) reflects semi-dynamic objects (red LiDAR points) and static objects (yellow LiDAR points). After classification, the weight of observation corresponding to semi-dynamic objects is reduced according to the distance they have moved. The decreasing ratio can be defined as $f(z_t, m_s, m_d, d_i^s, d_i^d)$. The result of the function is between zero and one, indicating how much the weight of current observation should be reduced. In this work, the positions of these semi-dynamic objects are supposed to follow a Gaussian distribution. Let F represent the decreasing ratio $f(z_t, m_s, m_d, d_i^s, d_i^d)$, which can be defined as:

$$F = \begin{cases} \exp\left(-\frac{(d_i^d)^2}{2\sigma^2}\right) & \text{if } \left|d_i^d - d_i^s\right| < \epsilon_1 \text{ and } d_i^d > \epsilon_2 \\ 1 & \text{otherwise} \end{cases} \tag{6.21}$$

where ϵ_1 is a relatively small distance, taking 0.1 in this section. ϵ_2 is the threshold at which an object is considered to be semi-dynamic, taking 0.3 in this section. Therefore, the final weight of current position can be computed as:

$$\omega = f\left(z_t, m_s, m_d, d_i^s, d_i^d\right) * p\left(z_t \mid x_t, m_s, m_d\right) \tag{6.22}$$

where $p(z_t|x_t, m_s, m_d)$ is the observation model. In this section, the likelihood observation model is chosen to calculate the weight of the current position. The pose estimation can be more accurate by using both the static map and semi-dynamic map. The algorithm of calculating the weight of the current pose is summarized in Algorithm 3.

One of the major advantages is that the three components in this framework, object detection, mapping, and localization algorithm, can be replaced by other corresponding mainstream algorithms.

Algorithm 3 Weight of pose

Input: Observation z_t, Static map m_s, Semi-dynamic map m_d;
Output: Weight of current pose;

$sig = 2 * sigma * sigma$
$weight = 0$
for each $lidarpoint$ **do**
  if $(z_i \neq z_{\max})$ **then**
    $x = pose\_x + range * cos\theta$
    $y = pose\_y + range * sin\theta$
    $dist_s = distance\ to\ nearest\ obstacles\ in\ m_s.$
    $dist_d = distance\ to\ nearest\ obstacles\ in\ m_d.$
    if $abs(dist_i^s - dist_i^d) < \epsilon_1$ and $dist_i^d > \epsilon_2$ **then**
      $coef = exp(-(dist_i^d)^2/sig)$
    else
      $coef = 1$
    end if
  end if
  $weight \mathrel{+}= coef * p\,(z_t \mid x_t, m_s, m_d)$
end for
return $weight$

6.3.2 *Experiment*

In order to validate the approach proposed in this section, experimental tests have been done using a real robot. The robot (Fig. 6.16) utilized in this work is equipped with a 16-line Velodyne LiDAR (only one beam is used during mapping and localization), a camera, an inertial measurement unit, and encoders.

The parking lot was selected as the experimental scenario because the parking lot changed significantly every individual run. It consists of four rooms, and the left

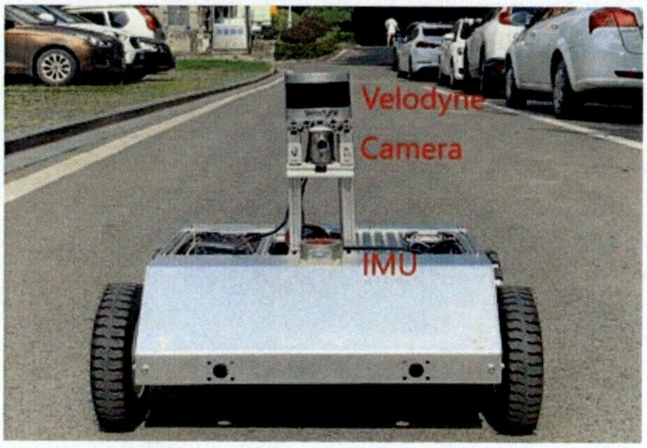

Fig. 6.16 The experimental platform

room in the map frame usually shows more cars compared with others. The robot started in the left room and went around the parking lot and back to its original place. Due to the particularity of the experimental scenarios, it is hard to directly get ground truth poses. We recorded the robot's poses during the mapping process as ground truth, and these poses during the localization process are also recorded using the map built on other days. The validation of proposed approach has been proved by comparing the localization accuracy without and with the proposed approach. The localization accuracy of using the proposed method in an altered environment was also compared to that of not using the proposed method in an unaltered environment.

For the mapping process, we steered the robot to collect data every few days. The data collected during the individual run were used to build m_s. m_d is constructed offline by using m_s and positions of semi-dynamic objects.

The final constructed m_s and m_d are shown in Fig. 6.17. In these static maps, some areas are distinct owing to the moving of cars, while the other parts are static all the time. m_d contains most of the semi-dynamic objects in the environment, and the missing semi-dynamic objects correspond to areas that are rarely observed, which have less impact on the localization accuracy. By applying the inverse observation model, the highly dynamic objects such as moving pedestrians and cars are automatically removed during the mapping process.

The localization accuracy with and without our algorithm in both altered and unaltered semi-dynamic environments is compared (Fig. 6.18a,b,c,d). In addition, we also present the localization accuracy of original Monte Carlo localization without our method utilizing a pure static map (Fig. 6.18e). All semi-dynamic objects, such as parked cars, were artificially removed. The poses during the mapping process were selected as ground truth, and the poses during the localization process were also recorded using maps built on other days. The algorithm Evo [13] was used to evaluate the accuracy of our approach offline. Figure 6.18 shows the qualitative result of localization accuracy, and Table 6.3 lists the quantitative comparison of the pose error. The MCL+− and MCL++ mean in an altered environment using Monte Carlo algorithm without and with our method (the first plus

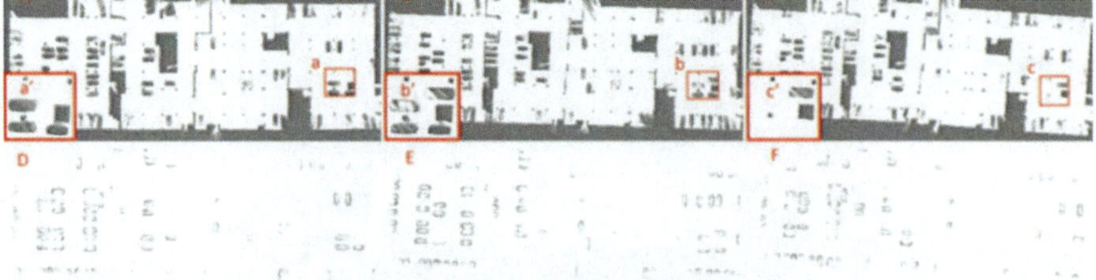

Fig. 6.17 The top panel (A, B, C) shows the static maps constructed using data collected in every individual run. (a', b', c') are the zoomed-in views of the area (a, b, c) where semi-dynamic objects (cars) change frequently. The bottom panel (D, E, F) shows the semi-dynamic maps constructed using information from static maps and semi-dynamic positions

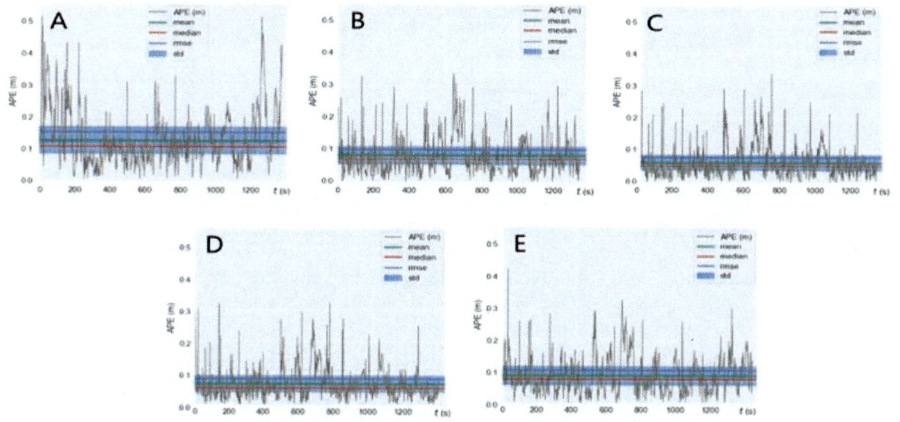

Fig. 6.18 Qualitative analysis of pose error. (**a**) and (**b**) show the localization accuracy without and with proposed method in an altered semi-dynamic environment. (**c**) and (**d**) show the localization accuracy without and with proposed method in an unaltered environment. (**e**) shows the localization accuracy of original Monte Carlo without proposed method using a pure static map

Table 6.3 Quantitative analysis of pose error

	MCL+−	MCL++	MCL−−	MCL−+	MCL*
Max [m]	0.516	0.336	0.336	0.327	0.423
Mean [m]	0.129	0.083	0.064	0.065	0.087
Min [m]	0.007	0.001	0.002	0.001	0.002
RMSE [m]	0.154	0.100	0.080	0.081	0.104
Std [m]	0.084	0.056	0.047	0.049	0.056

or minus subscript indicates whether the environment of the localization process has changed or not, and the second plus or minus subscript indicates whether the localization process uses our method or not). Therefore, the MCL−− and MCL−+ mean in an unaltered environment using Monte Carlo algorithm without and with our method. MCL* means using Monte Carlo algorithm without our method, using a pure static map. The pure static map is shown in Fig. 6.19 where the semi-dynamic objects are artificially removed from the map. Further, Fig. 6.18a,b,c,d,e correspond to MCL+−, MCL++, MCL−−, MCL−+, and MCL* in Table 6.3, respectively.

In Fig. 6.18a and b, our algorithm significantly decreases the deviation in altered environments, giving a lower mean error (0.083m versus 0.129m) and max error (0.336m versus 0.516m). According to the maps shown in Fig. 6.17 and the route of the robot, it encounters more semi-dynamic objects during the first 300 s and the last 200 s, and the localization accuracy fluctuates significantly in these places (Fig. 6.18a), which proves that the original Monte Carlo algorithm has low performance in altered environments. On the contrary, the results are more accurate after applying our algorithm (Fig. 6.18b).

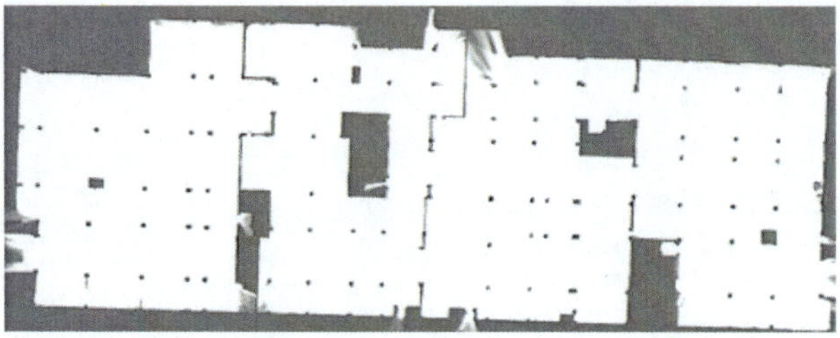

Fig. 6.19 The pure static map. Semi-dynamic objects are artificially removed

We also developed experiments to evaluate the algorithm performance in static environments. In Fig. 6.18c and d, our algorithm can also be applied in unaltered environments, giving relatively the same accuracy (0.327m versus 0.336, 0.065m versus 0.064) compared with traditional Monte Carlo localization method. That proves the versatility of the proposed methods in the environment with both static and semi-dynamic objects.

Besides, we also carried out experiments to test the localization performance of the original Monte Carlo localization algorithm in an altered environment, but the semi-dynamic objects are artificially removed from the map. Thus, the map only contains static objects of the environment (Fig. 6.19). The robot utilizes the static objects in the environment to localize the pose. The localization accuracy can be seen in Fig. 6.18e. It is reasonable that it has a lower accuracy as all the semi-dynamic information is discarded.

In addition, m_d is obtained offline. Therefore, during the localization process, the robot only needs to compare the distance between two maps, so the computational complexity is negligible. It should be noted that the absolute accuracy of the Monte Carlo algorithm is not the focus of this section. We aim to introduce our method into mainstream mapping and localization algorithms to obtain a more accurate result without sacrificing efficiency. And the experiment results have proven the better robustness, higher accuracy, and versatility of the proposed approach in both static and semi-dynamic environments.

During the process of getting semi-dynamic positions, there may be some semi-dynamic positions that hit the wrong place (Fig. 6.20a and b). However, the incorrect position does not have an influence on m_d generation as it has no corresponding connected area.

During the process of calculating the connected area, some static areas will be mistakenly regarded as semi-dynamic areas. In Fig. 6.20c, the pillar is mistaken for a semi-dynamic object; nevertheless, the static characteristic of these objects (pillar and wall) makes the weight of this area a constant. Therefore, the wrong connected area does not jeopardize the localization accuracy.

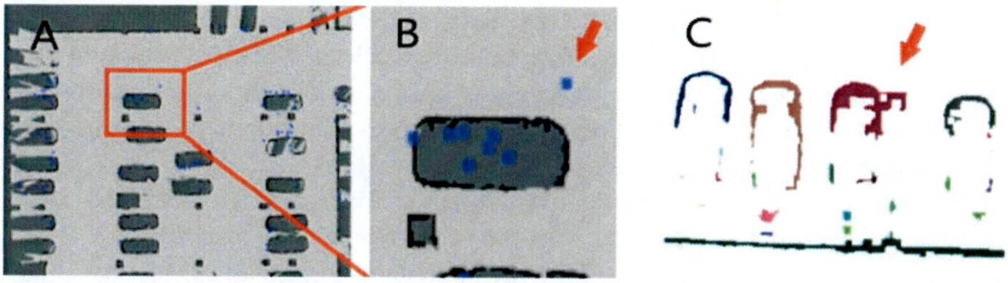

Fig. 6.20 Incorrect semi-dynamic position detection (**a** and **b**). (**b**) is the zoomed-in view of the red box area in (**a**). The blue dot pointed by the red arrow in (**b**) is incorrectly regarded as a semi-dynamic position. Incorrect connected areas (**c**). The place pointed by the red arrow in (**c**) is a pillar, which should be considered as a static object

6.4 Challenges and Prospect

Multi-sensor fusion technology has become the mainstream positioning solution for autonomous driving and robots. It has very broad prospects and also has a lot of space for development. The distribution of environmental features in the experimental scene we designed is relatively regular. If the attributes of obstacles in the scene are more complex, such as multiple obstacles stacking, occlusion, and irregular surfaces, these conditions may cause large errors in feature extraction, resulting in inaccurate map positioning. In addition, the pitch and vibration of the robot itself will bring the noise to the collected data, resulting in a decrease in accuracy. In future work, we will explore the integration of more complex environmental features, conduct experiments on more realistic roads, and try to add a variety of sensors, combined with deep learning to further improve the efficiency and quality of feature extraction, so that our model has generalization.

6.5 Summary

The environment feature distribution in the experimental scene we designed is relatively regular. If the obstacle attributes in the scene are more complex, such as multiple obstacles stacked, occluded, and irregular surfaces, these situations may cause large errors or even failures in feature extraction, resulting in map drift. Hence, we will explore integrating more complex environmental features in subsequent studies and also try to combine other sensors such as cameras to extract features to make our model universal.

In Sect. 6.2, a novel coarse-and-fine hybrid information fusion method was presented, GF-SLAM, for positioning under unstable external situations. Our system can real-time leverage the global information and feature-based SLAM while avoiding high computation costs and the influence of global information loss. By performing EKF positioning and building a feature map simultaneously when global information is accessible, we can achieve accurate positioning. However, when the global signal is lost, our system can smoothly switch to F-SLAM, which holds a basic localization precision. We believe our idea of hybrid-precision information fusion will inform continuing future research into the development of positioning and autonomous driving techniques.

In Sect. 6.3, we introduced a novel semantic map generation method and a localization method that can deal with the semi-dynamic environment. By fusing the camera and LiDAR, the presented method can autocratically detect and mark semi-dynamic objects during mapping. With this critical information, the 2D semi-dynamic map is constructed for later navigation, and a posterior distribution-based pose estimation is designed. We also apply several evaluations in an underground parking garage with a mobile platform equipped with a camera, LiDAR, and IMU. Results show that our method is able to work in most cases. Based on this work, our framework can be easily extended into 3D modes.

In a further extension, we will investigate the use of other types of maps and different localization algorithms, such as point cloud maps and NDT localization algorithms.

References

1. Aldibaja, M., Suganuma, N., Yoneda, K.: Robust intensity-based localization method for autonomous driving on snow–wet road surface. IEEE Trans. Ind. Inf. **13**(5), 2369–2378 (2017)
2. Andrade-Cetto, J., Sanfeliu, A.: Concurrent map building and localization on indoor dynamic environments. Int. J. Pattern Recognit. Artif. Intell. **16**(03), 361–374 (2002)
3. Aycard, O., Laroche, P., Charpillet, F.: Mobile robot localization in dynamic environments using places recognition. In: Proceedings. 1998 IEEE International Conference on Robotics and Automation (Cat. No. 98CH36146), vol. 4, pp. 3135–3140. IEEE (1998)
4. Cai, H., Hu, Z., Huang, G., Zhu, D., Su, X.: Integration of gps, monocular vision, and high definition (hd) map for accurate vehicle localization. Sensors **18**(10), 3270 (2018)
5. Chang, L., Niu, X., Liu, T., Tang, J., Qian, C.: Gnss/ins/lidar-slam integrated navigation system based on graph optimization. Remote Sens. **11**(9), 1009 (2019)
6. Chen, S.W., Nardari, G.V., Lee, E.S., Qu, C., Liu, X., Romero, R.A.F., Kumar, V.: Sloam: Semantic lidar odometry and mapping for forest inventory. IEEE Robot. Autom. Lett. **5**(2), 612–619 (2020)
7. Chiang, K.W., Tsai, G.J., Chang, H., Joly, C., Ei-Sheimy, N.: Seamless navigation and mapping using an ins/gnss/grid-based slam semi-tightly coupled integration scheme. Inf. Fusion **50**, 181–196 (2019)
8. Choi, M.J., Suhr, J.K., Choi, K., Jung, H.G.: Low-cost precise vehicle localization using lane endpoints and road signs for highway situations. IEEE Access **7**, 149,846–149,856 (2019)
9. Fan, Q., Sun, B., Sun, Y., Wu, Y., Zhuang, X.: Data fusion for indoor mobile robot positioning based on tightly coupled ins/uwb. J. Navig. **70**(5), 1079–1097 (2017)
10. Fox, D., Burgard, W., Thrun, S.: Markov localization for mobile robots in dynamic environments. J. Artif. Intell. Res. **11**, 391–427 (1999)
11. Gao, H., Zhu, J., Zhang, T., Xie, G., Kan, Z., Hao, Z., Liu, K.: Situational assessment for intelligent vehicles based on stochastic model and gaussian distributions in typical traffic scenarios. IEEE Trans. Syst. Man Cybern. Syst. **52**(3), 1426–1436 (2020)

12. Grisetti, G., Stachniss, C., Burgard, W.: Improved techniques for grid mapping with rao-blackwellized particle filters. IEEE Trans. Robot. **23**(1), 34–46 (2007)
13. Grupp, M.: evo: Python package for the evaluation of odometry and slam (2017). https://github.com/MichaelGrupp/evo
14. Guivant, J.E., Masson, F.R., Nebot, E.M.: Simultaneous localization and map building using natural features and absolute information. Robot. Auton. Syst. **40**(2-3), 79–90 (2002)
15. Guivant, J.E., Nebot, E.M.: Optimization of the simultaneous localization and map-building algorithm for real-time implementation. IEEE Trans. Robot. Autom. **17**(3), 242–257 (2001)
16. Hahnel, D., Triebel, R., Burgard, W., Thrun, S.: Map building with mobile robots in dynamic environments. In: 2003 IEEE International Conference on Robotics and Automation (Cat. No. 03CH37422), vol. 2, pp. 1557–1563. IEEE (2003)
17. He, G., Yuan, X., Zhuang, Y., Hu, H.: An integrated gnss/lidar-slam pose estimation framework for large-scale map building in partially gnss-denied environments. IEEE Trans. Instrum. Meas. **70**, 1–9 (2020)
18. Hess, W., Kohler, D., Rapp, H., Andor, D.: Real-time loop closure in 2d lidar slam. In: 2016 IEEE International Conference on Robotics and Automation (ICRA), pp. 1271–1278. IEEE (2016)
19. Hess, W., Kohler, D., Rapp, H., Andor, D.: Real-time loop closure in 2d lidar slam. In: 2016 IEEE International Conference on Robotics and Automation (ICRA), pp. 1271–1278. IEEE (2016)
20. Kohlbrecher, S., Von Stryk, O., Meyer, J., Klingauf, U.: A flexible and scalable slam system with full 3d motion estimation. In: 2011 IEEE International Symposium on Safety, Security, and Rescue Robotics, pp. 155–160. IEEE (2011)
21. Konolige, K., Grisetti, G., Kümmerle, R., Burgard, W., Limketkai, B., Vincent, R.: Efficient sparse pose adjustment for 2d mapping. In: 2010 IEEE/RSJ International Conference on Intelligent Robots and Systems, pp. 22–29. IEEE (2010)
22. Li, D., Gao, H.: A hardware platform framework for an intelligent vehicle based on a driving brain. Engineering **4**(4), 464–470 (2018)
23. Linegar, C., Churchill, W., Newman, P.: Work smart, not hard: Recalling relevant experiences for vast-scale but time-constrained localisation. In: 2015 IEEE International Conference on Robotics and Automation (ICRA), pp. 90–97. IEEE (2015)
24. Meyer-Delius, D., Hess, J., Grisetti, G., Burgard, W.: Temporary maps for robust localization in semi-static environments. In: 2010 IEEE/RSJ International Conference on Intelligent Robots and Systems, pp. 5750–5755. IEEE (2010)
25. Redmon, J., Farhadi, A.: Yolov3: an incremental improvement. Preprint (2018). arXiv:1804.02767
26. Ren, R., Fu, H., Wu, M.: Large-scale outdoor slam based on 2d lidar. Electronics **8**(6), 613 (2019)
27. Song, Y., Guan, M., Tay, W.P., Law, C.L., Wen, C.: Uwb/lidar fusion for cooperative range-only slam. In: 2019 International Conference on Robotics and Automation (ICRA), pp. 6568–6574. IEEE (2019)
28. Spangenberg, R., Goehring, D., Rojas, R.: Pole-based localization for autonomous vehicles in urban scenarios. In: 2016 IEEE/RSJ International Conference on Intelligent Robots and Systems (IROS), pp. 2161–2166. IEEE (2016)
29. Tardós, J.D., Neira, J., Newman, P.M., Leonard, J.J.: Robust mapping and localization in indoor environments using sonar data. Int. J. Robot. Res. **21**(4), 311–330 (2002)
30. Thrun, S., Burgard, W., Fox, D.: A probabilistic approach to concurrent mapping and localization for mobile robots. Auton. Robots **5**(3-4), 253–271 (1998)
31. Vineet, V., Miksik, O., Lidegaard, M., Nießner, M., Golodetz, S., Prisacariu, V.A., Kähler, O., Murray, D.W., Izadi, S., Pérez, P., et al.: Incremental dense semantic stereo fusion for large-scale semantic scene reconstruction. In: 2015 IEEE International Conference on Robotics and Automation (ICRA), pp. 75–82. IEEE (2015)
32. Wan, G., Yang, X., Cai, R., Li, H., Zhou, Y., Wang, H., Song, S.: Robust and precise vehicle localization based on multi-sensor fusion in diverse city scenes. In: 2018 IEEE International Conference on Robotics and Automation (ICRA), pp. 4670–4677. IEEE (2018)

33. Wang, B.H., Chao, W.L., Wang, Y., Hariharan, B., Weinberger, K.Q., Campbell, M.: Ldls: 3-d object segmentation through label diffusion from 2-d images. IEEE Robot. Autom. Lett. **4**(3), 2902–2909 (2019)
34. Wolf, D.F., Sukhatme, G.S.: Mobile robot simultaneous localization and mapping in dynamic environments. Auton. Robots **19**(1), 53–65 (2005)
35. Xu, L., Feng, C., Kamat, V.R., Menassa, C.C.: An occupancy grid mapping enhanced visual slam for real-time locating applications in indoor gps-denied environments. Autom. Constr. **104**, 230–245 (2019)
36. Yu, C., Liu, Z., Liu, X.J., Xie, F., Yang, Y., Wei, Q., Fei, Q.: Ds-slam: a semantic visual slam towards dynamic environments. In: 2018 IEEE/RSJ International Conference on Intelligent Robots and Systems (IROS), pp. 1168–1174. IEEE (2018)

Part III
Advance

This part of the book is the frontier exploration and extension of multimodal fusion perception technology. It consists of three parts. Chapter 7 presents a dataset for autonomous driving perception dedicated to the development of multimodal fusion algorithms. This dataset is a Chinese city dataset established to fill the research and development gap. In Chap. 8, the fusion technology is no longer limited to the vehicle itself, but extends to the collaboration between the vehicle and the road. Finally, the research on the underlying mechanism of multimodal fusion is in Chap. 9, providing a solution to enhance the fusion effect.

Chapter 7
OpenMPD: An Open Multimodal Perception Dataset

Abstract Multimodal sensor fusion techniques have promoted the development of autonomous driving, while perception in a complex environment remains a challenging problem. This chapter proposes the Open Multimodal Perception Dataset (OpenMPD), a multimodal perception benchmark aimed at difficult examples. Compared with existing datasets, OpenMPD focuses more on those complex traffic scenes in urban areas with overexposure or darkness, crowded environments, unstructured roads, and intersections. It acquires the multimodal data through a vehicle with 6 cameras and 4 LiDAR for a 360-degree field of view and collects 180 clips of 20-s synchronized images at 20Hz and point clouds at 10Hz. In particular, we applied a 128-beam LiDAR to provide Hi-Res point clouds to understand the 3D environment and sensor fusion better. We sampled 15K keyframes at equal intervals from clips for annotations, including 2D/3D object detections, 3D object tracking, and 2D semantic segmentation. Moreover, we provide four benchmarks for all tasks to evaluate algorithms and conduct extensive 2D/3D detection and segmentation experiments on OpenMPD. Data and further information are available at http://www.openmpd.com/.

7.1 Introduction

During the past decades, autonomous driving technology has made great progress in localization, perception, planning, control, etc. Most of the problems in typical situations can be solved with existing algorithms, while those complex corner cases are still unsolved. Therefore, we propose a new dataset, OpenMPD, to promote the development of research [7, 28].

X. Zhang et al., *Multi-sensor Fusion for Autonomous Driving*,
https://doi.org/10.1007/978-981-99-3280-1_7

Before that, different groups have put out different sets of data to speed up the development of the field of self-driving cars [3, 14, 27, 37, 45]. Since the Karlsruhe Institute of Technology (KIT) and Toyota Technological Institute at Chicago (TTIC) provided KITTI [12] as datasets in 2012, more complementary datasets have been produced along with the upgrading of autonomous vehicle techniques. Autonomous driving perception can be divided into single-modal and multimodal algorithms. Single-mode methods generally only use 3D point clouds or visual images. They suffer from the limitations of complex scenes like night and fog (e.g., [33, 34, 38]). Therefore, multimodal algorithms are getting more and more attention, but there are some challenges in multimodal data fusion, which lie in the imbalance of data space and feature space between different modalities. At the same time, multimodal algorithms often require mass computation. Different tasks may need different annotations, so the combination of multi-task (simultaneously using object detection and semantic segmentation fusion) and multimodal (simultaneously using multi-sensor fusion) has become a trend [15].

Common autonomous driving datasets already contain visual and LiDAR information, but few of them provide all the annotation for different tasks. Under these circumstances, we provide an OpenMPD as shown in Fig. 7.1, which has 2D bounding box annotation, 3D bounding box annotation, 2D semantic segmentation, and 3D semantic segmentation in each keyframe. We divide the driving environment into seven categories, mainly for complex road conditions. They are main roads, two-way single lanes, irregular roads, intersections, tunnels/culverts, internal roads (e.g., university campuses and communities), and different heights (e.g., viaduct). Some typical examples are shown in Fig. 7.2. At the same time, we divide driving scenes for the poor performance of sensors into six categories: light and shadow change, road construction/traffic accident, U-turn, night, dense objects, and object occlusion. In addition, we have more crowded and occluded scenes, and the average number of pedestrians and vehicles per frame is higher than in other datasets, as shown in Fig. 7.3.

We point out that advanced sensors can benefit research, especially for LiDAR. Existing datasets use sparse point clouds. The density of point clouds can be improved by point cloud completion methods, but it may introduce deviation. In order to obtain dense point clouds, we used 128-beam LiDAR in the OpenMPD for the first time, supplemented by a 40-beam and two 16-beam LiDARs. Dense point cloud is very important for depth map and high-precision mapping [1, 16, 20]. Furthermore, we use six cameras to achieve 360-degree shooting, which is very important for multi-view and other tasks.

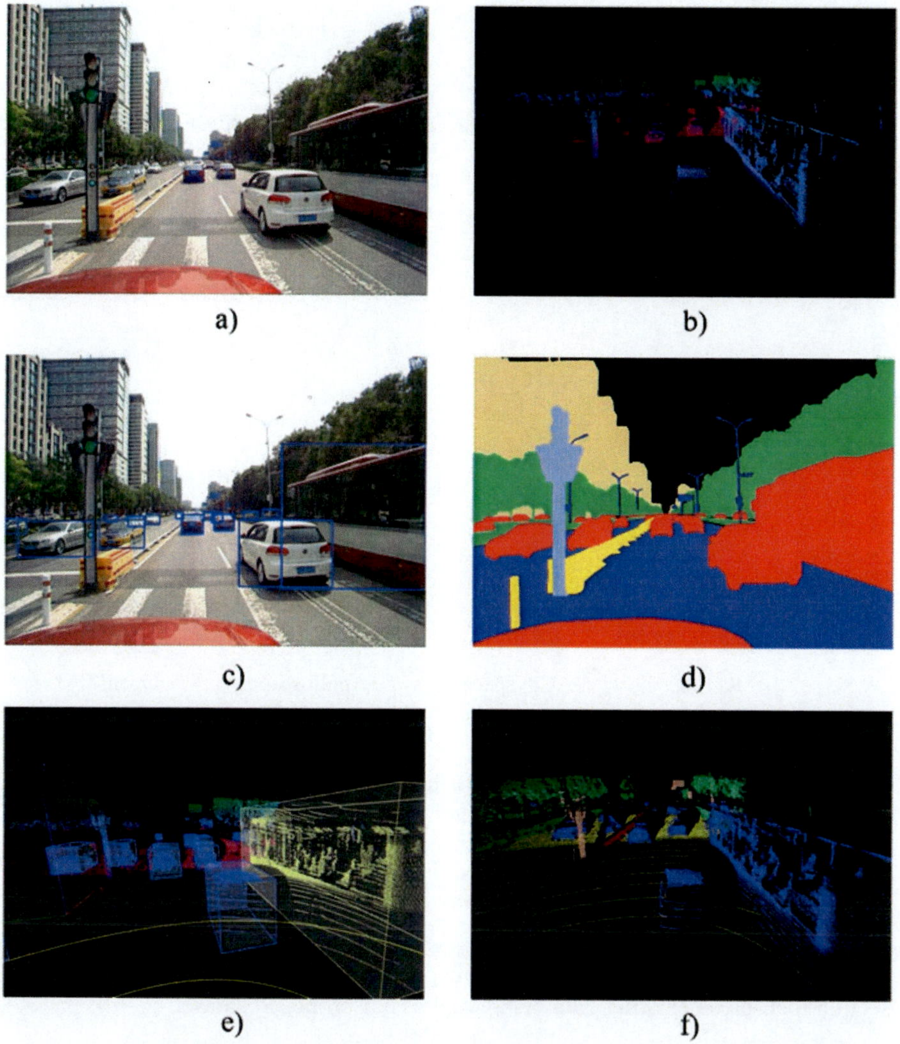

Fig. 7.1 Examples of three types of annotation. (**a**) represents the original image captured by the camera; (**b**) represents the image displayed by LiDAR; (**c**) represents the 2D bounding box annotation on the RGB image; (**d**) represents the 2D semantic segmentation annotation on the RGB image; (**e**) represents the 3D bounding box annotation in the point cloud; and (**f**) represents the 3D semantic segmentation annotation in the point cloud

Fig. 7.2 Some typical examples in OpenMPD

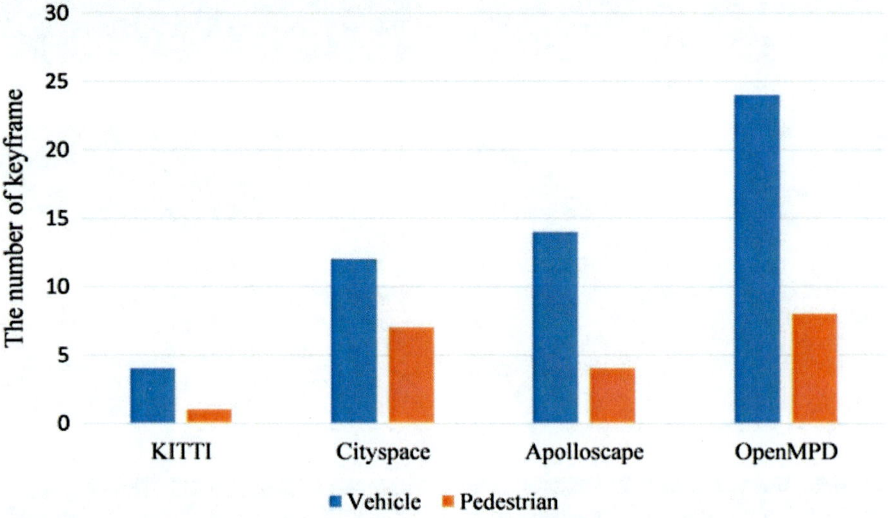

Fig. 7.3 Comparison of the complexity of different datasets. The number of each 2D column represents the average number of vehicles and pedestrians in the dataset keyframes

7.2 Automated Driving-Related Datasets

Datasets are the basis for developing perceptions for autonomous driving. This subsection focuses on existing datasets as well as our proposed dataset.

7.2.1 Comprehensive Datasets

As one of the earliest multimodal datasets, KITTI [12] contains point clouds from LiDAR, front-facing images from two kinds of cameras, and GPS/IMU data. This dataset also contains complex scenes since it is collected from city, urban, and highway scenes. It provides both 2D and 3D annotation data, and the number of both 2D bounding boxes and 3D bounding boxes is about 80k. However, this dataset is limited by its 2D part size, equipment, and weather. It contains only about 15k RGB images and point clouds. However, all of its data is collected on sunny days. Environments like snowy days and rainy days are not considered.

As one of the largest multimodal datasets, nuScenes [4] contains 1.4 million images and 400K point clouds. It has a real 360-degree collecting sensor built by six cameras, one LiDAR, and five RaDARs. Compared with LiDAR, the accuracy and density of RaDAR are poor. In this situation, it selected a 32-beam LiDAR, which has a short range but provides dense point clouds compared with RaDAR. However, it only provides 3D bounding box annotation, which is very unfriendly for multi-task learning.

As one of the largest and most diverse multimodal autonomous driving datasets, Waymo [32] has five high-resolution cameras and five high-quality LiDARs, and the data that they collect are dense and clear. They provide 12 million 3D bounding boxes and 9.9 million 2D bounding boxes from its 1150 scenes. In addition, its detection and tracking mainly rely on LiDAR rather than cameras. However, the angle of view of the camera is less than 270 degrees.

ApolloScape [17], released in 2017, contains 143,906 image frames. It not only provides three-level difficulty settings for users to test and train algorithms but also designs an efficient 2D/3D joint marking pipeline to save about 70% of the marking time. However, the laser scanner has lower precision in collecting data compared to LiDAR.

A2D2 [13] is built on a 360-degree full-coverage sensor platform, which contains six cameras and five LiDAR units. With this sensor platform, they provide 3D bonding box annotation, point cloud segmentation, and semantic segmentation.

They mainly focus on semantic segmentation, but the number of their annotations is limited in size compared with other newly proposed datasets. The perception system of 16-beam sparse LiDAR and the lack of 2D bounding box annotation limit the performance and adaptability of the dataset.

7.2.2 Characteristic Datasets

TuSimple provided a set of more than 1000 2-s-long video clips. The speed and position are generated by the vehicle's distance sensor in the last frame. It also provided human-annotated bounding boxes as supplementary training data for vehicles on more than 5000 images. In addition to training and supplementary data, it also provided external data sources to help the algorithm learn.

CULane [30] was a challenging large traffic lane detection dataset, which used cameras to collect data on six different cars in Beijing. The dataset collected more than 55 h of video and extracted 133,235 frames from the video. For each frame, the lane is manually marked with a cubic spline curve.

Different annotations may have different effects on different scenes, and none of them can be completely ignored. As a result, every dataset creator selects the annotation that they think is the best [6, 26, 35, 42, 43]. For example, nuScenes [4] annotates RGB images by using 3D bounding boxes, while it annotates point clouds by using 2D bounding boxes. Waymo [32] annotates RGB images by using 2D bounding box. However, multimodal and multi-task algorithms require synchronous annotation on RGB images and point cloud.

7.2.3 *Our Dataset*

Due to the sparse point cloud, some 3D datasets contain few or small objects, which makes it difficult to carry out a large number of experiments [22, 24, 25, 29, 36, 39]. Aiming at the requirements of multi-view, multi-task, and multi-mode fusion, we propose OpenMPD Benchmark. In Table 7.1, OpenMPD not only uses six cameras to build a 360-degree view but also expands the sensing coverage through the 128-beam LiDAR. Our dataset is more in line with real life, containing seven complex driving environments and six driving scenes as described in the introduction. Each selected frame contains six types of annotations, including RGB images and point cloud modalities.

Table 7.1 Comparison with the state-of-the-art datasets

	RGB image	Point cloud	Degree of camera view	Point cloud beams	Year
nuScenes [4]	3D bounding box	2D bounding box	360	32	2020
H3D [31]		3D bounding box	180	64	2019
ApolloScape [17]	2D semantic segmentation		360		2019
A2D2 [13]	2D semantic segmentation	3D semantic segmentation	270	5*16	2020
	3D bounding box				
Waymo Open [32]	2D bounding box	3D bounding box	270	75, 4*20	2020
Level 5 Lyft	2D bounding box	3D bounding box	360	64, 2*40	2019
KITTI [12]	3D bounding box	2D bird eye view	180	64	2012
	2D bounding box				
OpenMPD	2D bounding box	2D bounding box	360	128, 40, 2*16	2021
	2D semantic segmentation	3D bounding box			
	3D bounding box	3D semantic segmentation			

7.3 OpenMPD

In this chapter, we mainly introduce the data acquisition equipment and the details of annotation. What's more, we also introduce the sensor calibration and data detailed acquisition plan.

7.3.1 Platform Setup

The OpenMPD data acquisition platform is shown in Fig. 7.4. Six cameras and four LiDARs are installed on the rooftop of the car. On the front side of the rooftop of the car, there are three cameras and a 128-beam LiDAR. We also placed a camera and a LiDAR on the left, right, and rear sides of the rooftop of the car, respectively. The explicit configuration of the camera is shown in Fig. 7.5a. Specifically, two cameras on the left and right sides of the rooftop of the vehicle with a 6 mm focal length form a binocular vision through calibration. The two cameras are jointly responsible for collecting objects with a horizontal viewing angle of 110 degrees within 50 m. The middle camera, with a 25 mm telephoto lens, is responsible for collecting objects with a horizontal angle of view of 23 degrees and 50–150 m. Two 5 mm wide-angle cameras are responsible for collecting the horizontal angle and placing the joined objects. The rear side is equipped with an 8-mm wide-angle camera, which is responsible for collecting objects with a horizontal viewing angle of 100 m. All cameras are uniformly set to a resolution of 1024×968, using a complementary metal-oxide semiconductor (CMOS) sensor at a frame rate of 20Hz.

The LiDAR with 128 beams located at the front rooftop of the vehicle is used as the main scene acquisition device to provide a 360-degree coverage of the whole environment. The explicit configuration of LiDAR is shown in Fig. 7.5b. Compared

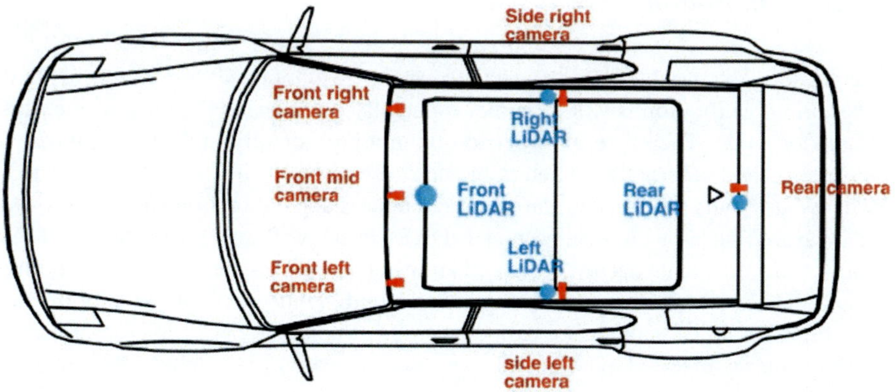

Fig. 7.4 The red dot represents the camera, and the blue dot represents the LiDAR

(a) The detailed configuration of six cameras in different positions of the vehicle.

Sensors parameter	Number	Resolution	Focal length	Vertical FOV	Data transfer speed	Range
Front middle	1	1024*968	25mm	23°	47.41Mbp/s	50-150m
Front Left And Right	2	1024*968	6mm	100°	47.41Mbp/s	50m
Behind	1	1024*968	8mm	65°	44.40Mbp/s	100m
Side Left And Right	2	1024*968	5mm	78°	16.41Mbp/s	3m

(b) The detailed configuration of four LiDAR in different positions of the vehicle.

Sensors parameter	Number	Beam	Vertical FOV	Vertical FOV	Range(official)	Range(tested)
Front	1	128	-25°~15°	0.1	200m	150m
Behind	1	40	-25°~15°	0.33	200m	100m
Side Left And Right	2	16	-15°~15°	2	100m	30m

Fig. 7.5 The detailed configuration of sensors in the data acquisition platform

to those LiDARs with fewer beams, it can collect information from a further distance of up to 150 m away and obtain more accurate data and a denser number of points. As a supplement to the LiDAR visual blind area, we use two 16-beam LiDARs on two sides and one 40-beam LiDAR on the back to cover the blind spot. All LiDARs are synchronized at a rate of 10Hz.

The OpenMPD dataset was collected from October 2020 to November 2020. Our data collection system activates all the sensors and records data in ROS Kinetic Kame system based on Ubuntu 16.04. The ROS system can subscribe to the topics of different sensors separately and receive multiple topics uniformly through a Time Synchronizer. Only when all topics have the same time stamp will a callback function of the synchronization result be generated to process the data after the synchronization time. In this way, the timestamp can be synchronized, and the picture can be paired with its corresponding point clouds.

7.3.2 *Calibration*

The sensors are mounted on the vehicle, and measurements were made of their poses. Field calibration is still required, especially considering the error of measuring sensor angle. Therefore, in order to obtain a high-quality multi-sensor dataset, the internal and external parameters of the sensor need to be carefully calibrated. Here, we use tools such as calibration object boards and calibration spheres. Firstly, the measured pose of the front-center LiDAR sensor with respect to the reference frame is assumed to be accurate. All remaining sensor poses are determined relative to this LiDAR, which serves as a reference. Secondly, intrinsic camera calibration is performed for all cameras using calibration object boards (checkerboards). Finally, in determining external parameters, we use the calibration sphere and mainly calculate the center of the calibration sphere from the LiDAR point cloud and the image. In the LiDAR point cloud, the center of the ellipse is steadily determined using random sample consensus (RANSAC). In the image, the contour of the sphere is detected by the edge, and then the ellipse is fitted to the edge point. Since the size of the ellipse is known according to the ellipse parameters, the center of the sphere is also determined.

In determining the center of the sphere, we mainly take the following steps:

1. Run the Calibration tools and play data.
2. Image: The color can be selected through the hue, saturation, and value (HSV) range and adjusted according to the recognition effect.
3. LiDAR: The range of candidate point clouds can be reduced by adjusting their x, y, z range and by adjusting the sphere radius R. So we can adjust to get the right recognition effect.
4. We project the point cloud onto the 2D plane of the image and judge the similarity score between the point cloud projection and the image. When it is greater than a certain threshold, we consider the point cloud and the image to perform well and record the center point of the sphere. We record the data in different positions several times to calculate the external parameters.

The result of the calibration is given in a configuration file containing:

- The following parameters for each camera sensor: $ImageSize \in N^2$, resolution (columns, rows) of camera image; $RadialDistortion \in R^2$, radial distortion parameters in the camera; and $IntrinsicMatrix \in R^{3*3}$, the intrinsic camera matrix of the camera image.
- The following parameters for the external parameters of the LiDAR and camera pair: $R \in R^{3*3}$, represent the rotation matrix, and $T \in R^{3*1}$, represent the translation matrix.

7.3.3 Collecting Route

In order to create the complexity of the driving scenes, we select the specific weather conditions and the section of the road that can provide us with the opportunity to collect those extreme and diverse environments for sensors. In the summer, we use a sport utility vehicle (SUV) to collect data at a speed of 40km/h in urban areas of Beijing, China. During this period, we started from the Haidian District and went all the way to the suburbs and even the countryside to collect more diverse data. Besides, we collect complex data on university campuses, highways, suburban villages, parking lots, etc. Considering a variety of road conditions, we also collect complex scenes in viaducts, tunnels, turntable roads, and other places. The data collection started in the middle of October 2020 and lasted for 1 month. In the bounding box, there are more than 228,000 objects on OpenMPD. For semantic segmentation, there are more than 550,000 objects on OpenMPD. It will provide a complex and diverse dataset for the autonomous driving algorithm, which is of great significance to testing the robustness of the model and the real vehicle test.

7.3.4 Combine Annotation

We annotate our dataset with detection and segmentation labels in both images and point clouds. For the selected RGB images, we annotate them with 3D bounding boxes, 2D bounding boxes, and 2D semantic segmentation. For point clouds, we provide annotation in 3D bounding boxes, 2D bounding boxes, and 3D semantic segmentation. Six annotations of two modalities are combined to make the dataset available for various multimodal and multi-task algorithms.

In order to collect the data reflecting real, complex situations in the real world, first, we collect raw data from those places where we believe we can cause a problem for the autonomous driving algorithm. Then, the annotation data will be selected manually from the raw data and partitioned every 20s. Those data with difficult environment (e.g., main road, two-way single lane, irregular road, intersection, tunnels/culvert, internal road, and different heights) or complex driving scenes (e.g., road construction, accidents) will be seen as the selected data. For both 2D annotation and segmentation, we followed these steps. We sample keyframes at 1Hz for each scene, with a pair of images and point clouds that are synchronized. All keyframes are annotated with different kinds of bounding box and semantic segmentation.

Our annotation method consists of two parts. One is annotation with bounding box, which includes all kinds of annotation with 2D bounding box and 3D bounding box, and the other is annotation with semantic segmentation, which includes semantic segmentation on both RGB images and point clouds.

We selected 15,000 RGB images and point clouds. For the bounding box, there are more than 228,000 objects on OpenMPD. We annotated six classes of objects, including cars, people, mini-vehicles, trucks, buses, and engineering vehicles. For semantic segmentation, there are more than 550,000 objects on OpenMPD. The categories for semantic segmentation pay more attention to the road situation for autonomous driving, including fence, lane line, background, and so on, as shown in Table 7.4. Lane, sidewalk, crosswalk, and lane line give the information of right path to run on for the algorithm, and traffic light and diversion line help the algorithm to follow the traffic rule and do the right decision. We hope this dataset can contribute to the development of autonomous driving technology.

7.4 Data Analysis

In this chapter, we analyze the complexity, occlusion, scale, and location of the OpenMPD dataset.

7.4.1 Complexity

Autonomous driving technology has made great progress and is applied to most scenes, but what hinders its further development are complex scenes. The complex dataset required quantitative analysis, which depends on not only the number of objects in each keyframe but also the complex driving environment (road conditions) and driving scenes (traffic accidents and climate change). On the other hand, object occlusion is also an important factor in judging whether the dataset is complex. Furthermore, the type of driving position also has a strong impact on driving. For example, driving on a bridge with a high angle of view can easily affect the perception module of the vehicle. To evaluate the complexity of a dataset, we have set up several key factors for both scenes and environments. The complexity of one dataset is shown based on the average number of key factors in a frame.

We set up seven factors to determine the complexity of the driving environment. They are main roads, two-way single lane, irregular roads, intersections, tunnels/culverts, internal roads, and different heights (e.g., viaduct). Based on the statistics in Table 7.2, we find that 57.92% of the scenes that we selected include intersections and two-way single lane occupied 43.16% of all of the selected senses. These two scenarios are intuitively more complex than straight trunk roads. In different situations, the autonomous driving algorithm needs to face different challenges. While driving on the main road, the algorithm needs to deal with crowded and faster traffic. On the other hand, the two-way single lane provided the vehicle less space to drive on. More comprehensive driving environment data puts forward comprehensive requirements for autonomous vehicles.

As shown in Table 7.3, we also collected six driving scenes that can be considered as unexpected cases. They contain light/shadow changes, road construction/traffic

Table 7.2 Seven driving environments and the corresponding appearance rate of the dataset. A frame image may contain multiple scenes at the same time

Factors of environment	Rate
Main road	36.06%
Two-way single lane	43.16%
Irregular road	20.21%
Intersection	57.92%
Tunnels/culvert	9.29%
Internal road	29.50%
Different heights	8.19%

Table 7.3 Six driving scenes and the corresponding appearance rate of the dataset. A frame image may contain multiple scenes at the same time

Factors of scenes	Rate
U-turn	34.42%
Night/dark place	10.38%
Dense objects	56.83%
Object occlusion	90.16%
Light/shadow change	30.60%
Road construction/traffic accident	21.85%

accidents, U-turn, night, dense objects, and object occlusion by overtake. It is difficult for us to directly measure the impact of a scene, so we use the number of scenes as an indicator. We use the number of unexpected cases to figure out the complexity level of driving scenes. Therefore, we classify scenes to facilitate model learning and verify the model's performance in different scenes.

In terms of bounding box, we define six categories for vehicle and pedestrian as shown in Table 7.4. Each keyframe contains an average of 17 vehicles and 6 pedestrians. For crowded scenes, the number of vehicles may reach 25, and the number of pedestrians can reach more than 20, which is higher than most of the modern datasets. On the other hand, there are dark places like streets without lamps, and they will have fewer objects that are necessary for our dataset.

In terms of segmentation, we used 11 categories to annotate all driving signs, as shown in Table 7.5. For segmentation, each keyframe contains about 37 annotation objects on average. In addition, we focus on scenes on autonomous driving roads, such as lanes, lane lines, and lane markings. They are a challenge for existing semantic segmentation methods, but they are very important for autonomous driving. The full details of road information will help the vehicle make the right decision during a sudden change in information from sensors.

After analyzing the complexity of our dataset, we believe this dataset presents a complex and challenging environment for the autonomous driving field.

Table 7.4 Six classes of object detection and rate of appearance

2D classes	Rate	3D classes	Rate
Car	54.96%	Car	60.55%
Person	34.57%	Person	18.55%
Mini-car	3.10%	Mini-car	12.40%
Truck	2.56%	Truck	1.56%
Bus	3.06%	Bus	2.44%
Engineering vehicle	1.75%	Engineering vehicle	4.5%

Table 7.5 Eleven classes of object detection and rate of appearance

2D classes	Rate	3D classes	Rate
Fence	3.56%	Fence	1.51%
Road	26.36%	Road	24.91%
Background	2.40%	Background	0.81%
Traffic sign	14.88%	Traffic sign	1.20%
Lane line	14.40%	Lane line	11.84
Sidewalk	9.23%	Sidewalk	2.81%
Lane marking	5.14%	Green plants	21.18%
Crosswalk and diversion line	4.74%	Building	28.93%
Traffic light	5.99%	Cyclist	0.39%
Person	7.17%	Person	0.48%
Car	6.13%	Car	5.94%

7.4.2 Occlusion

Occlusion brings challenges to detection and segmentation, especially for severe occlusion. In rush hours, there is occlusion between pedestrians and vehicles. Considering the depth information of the camera, objects closer to the camera may block farther away from the camera. Firstly, we divide the level of occlusion and define it by the intersection of unions (IoU) of different objects. If the IOU is 0, it means not occluded. If the IOU is 0 to 0.5, it indicates partial occlusion. More than 0.5 is considered as severe occlusion. According to our statistics, more than 29.07% of the objects are heavily occluded, 51.48% are partially occluded, and the rest are not occluded.

7.4.3 Scale

Since the car and the person are the most common participants, we are going to analyze the size of their bounding boxes to evaluate the driving conditions at a safe distance. And safe distance is affected by driving speed. Higher driving speeds need longer distances to brake. The scale of an object represents the distance from us to this object. To determine the scale, we simply look at the height of each bounding box, like what KAIST [18] does, since areas can always be affected by both the length and the width of the object. After our calculation, we find out that the medium height of our person is 43 and the medium height of our car is 47. Furthermore, the domain for our person is 5 to 180, and that for our car is 5 to 200. With 40 pixels as high, a 1.75 m person shown in an image is about 30m away. We assume that the average speed of cars that travel in the city of Beijing is 35 km/h to 60 km/h and that the distance for a car to brake is about 23m to 45m. Most of the objects in front should be no closer than 23m to 45m. For the perception of the expressway, this distance can be increased to 115 m.

After that, we found out that the height of my car and the person in it show the same pattern in the image. As shown in Figs. 7.6 and 7.7, most of our objects are small, no matter if it is an image of a car or a person. It proves the high density of the object distribution of our dataset. In a crowded environment, most nearby objects will have larger sizes, so they are easily covered by each other.

7.4.4 Position

Knowing the position of most objects helps the algorithm determine the area of focus. In order to figure out the position attribution of objects in this dataset, we split all the objects into two parts. One is their distance, which represents how far

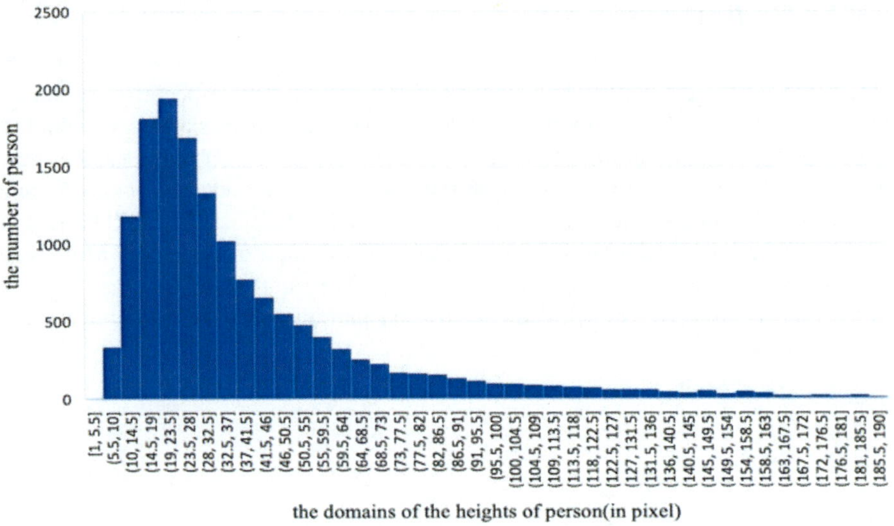

Fig. 7.6 The histogram of person height distribution

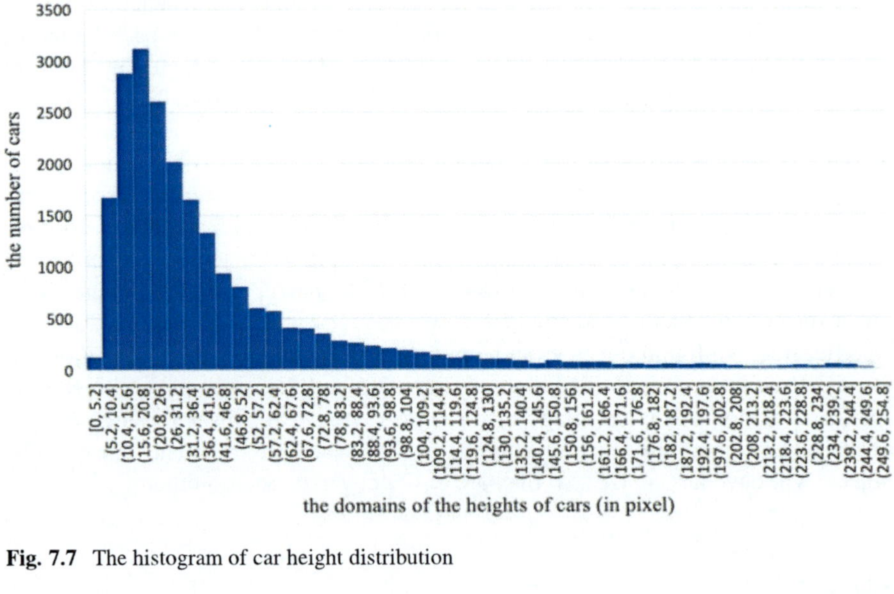

Fig. 7.7 The histogram of car height distribution

away an object is from the camera, while another is their location, which shows whether they are in the same lane as the vehicle or not.

The detection method can calculate the anchor box by modeling the distance between the object and the camera. The height of an object's bounding box can always represent the size of this object. As we mentioned in the scale part, compared to the area of an object's bounding box, analyzing the height of an object's bounding box can be more reliable and more reasonable since the factors that could affect the height of an object's bounding box are only its original height, the angle of vision, and how far the vehicle is from it. There are two factors that matter to the height of an object's bounding box, which are distance and the occupation of visual area. Therefore, as a pedestrian is walking far away, both the distance and the occupation of the visual area change. Considering the safe distance that we mentioned on the scale part, we need to find out the height of a pedestrian in pixels at a distance of 23 to 45m. After our measurement, it will be about 54 pixels to 27 pixels. As shown in Fig. 7.8, most of our pedestrians are less than 30 pixels, and they will be at a safe distance from vehicles in the city.

To evaluate the location of most objects, we only separate the lane by two parts. One is in the same lane, while the other is on the side. It is easy to use our semantic segmentation annotations to find the zone of our vehicle lane. We regard other vehicles and pedestrians moving in the same lane as we are currently driving in as the same lane. We need to be more cautious about those scenes that are not in the same lane, which causes many complex scenes. Overtaking, occupying another lane, doing a U-turn, and changing lanes are more challenging than following a vehicle on the lane. After our calculation, the number of objects that are in the same lane as

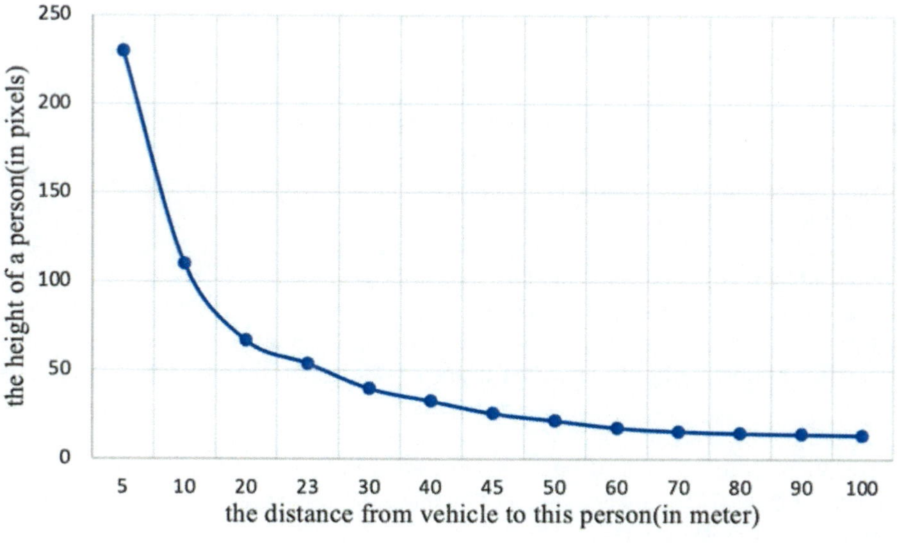

Fig. 7.8 Relationship between pedestrian height and distance from camera on vehicle

our vehicle is only 3.6305%. The rest of the objects are things that we need to put more focus on.

7.5 Experiment

In this chapter, we introduce the benchmark test results on the OpenMPD dataset, including 2D/3D object detection and 2D/3D semantic segmentation.

7.5.1 Object Detection

For 2D object detection, we evaluate it using mean average precision (mAP) followed by the COCO dataset [23]. We divide the dataset into a training set, a validation set, and a test set according to 8:1:1. The experiments aim at six classes of interest (shown in Table 7.4) and one background class.

We use YOLOv5 [19], YOLOX [11], and FCOS [41] methods to perform model training on OpenMPD by loading pre-training weights from the COCO dataset. For YOLOv5 and YOLOX, we have selected two models, one with faster speed and the other with better accuracy. The experimental results are shown in Table 7.6, where speed, parameters, and flops represent the inference speed, network parameters, and calculation of the model, respectively. The mAP@.5 indicates the average AP of each class when IoU is set to 0.5, while mAP@[.5, .95] represents the mean average precision from IoU threshold from 0.5 to 0.95 in steps of 0.05. The mAP@.5 of most methods is higher on COCO, and mAP@[.5, .95] is higher on OpenMPD from Table 7.6. In addition, OpenMPD has only 6 classes for applications in the field of autonomous driving, but COCO has 80 classes, which is an important reason for the low mAP@[.5, .95] on COCO. Figure 7.9 shows the prediction results of different methods. It can be seen intuitively that there are a large number of objects and partial occlusion. The last two columns of Fig. 7.9 show the low-light scene of OpenMPD when the car passes through the tunnel.

For 3D object detection, we evaluate it using mean average precision (mAP) followed by KITTI [12]. We annotate six categories as shown in Table 7.4, which is different from the three classes of KITTI, e.g., car, pedestrian, and cyclist. The size of different classes of vehicles is very vital for autonomous driving, so we classify vehicles in more detail. We used SECOND [44], PointPillars [21], and TANet to train and test on OpenMPD. The experimental results are shown in Table 7.7. KITTI has 3 classes, and OpenMPD has 11 classes, which is more conducive to the perception of autonomous driving. Compared with the mAP of the corresponding method, OpenMPD is lower than KITTI, which also reflects that OpenMPD is more oriented to complex scenes.

Table 7.6 Comparison of 2D object detection methods on OpenMPD and COCO datasets. The speed, parameters, and flops are tested on a single RTX 3090. "–" denotes the corresponding information is not reported in the chapter. "↑" means higher score is better

Model	OpenMPD		COCO		Speed (ms) ↓	Parameters (M) ↓	FLOPs(G) ↓
	mAP@.5(%) ↑	mAP@[.5, .95](%) ↑	mAP@.5(%) ↑	mAP@[.5, .95](%) ↑			
YOLOv5-s [19]	60.7	37.8	56.0	37.2	**3.5**	7.1	**8.2**
YOLOv5-x [19]	43.6	63.6	68.9	50.7	13.1	87.2	217.2
YOLOX-s [11]	**63.7**	41.2	–	40.5	16.5	8.8	13.3
YOLOX-x [11]	46.3	**69.2**	**69.6**	**51.1**	60.8	99.0	140.7
FCOS [41]	52.6	31.8	57.4	38.6	50.0	**5.9**	86.2

Bold values highlight the significant improvement in performance

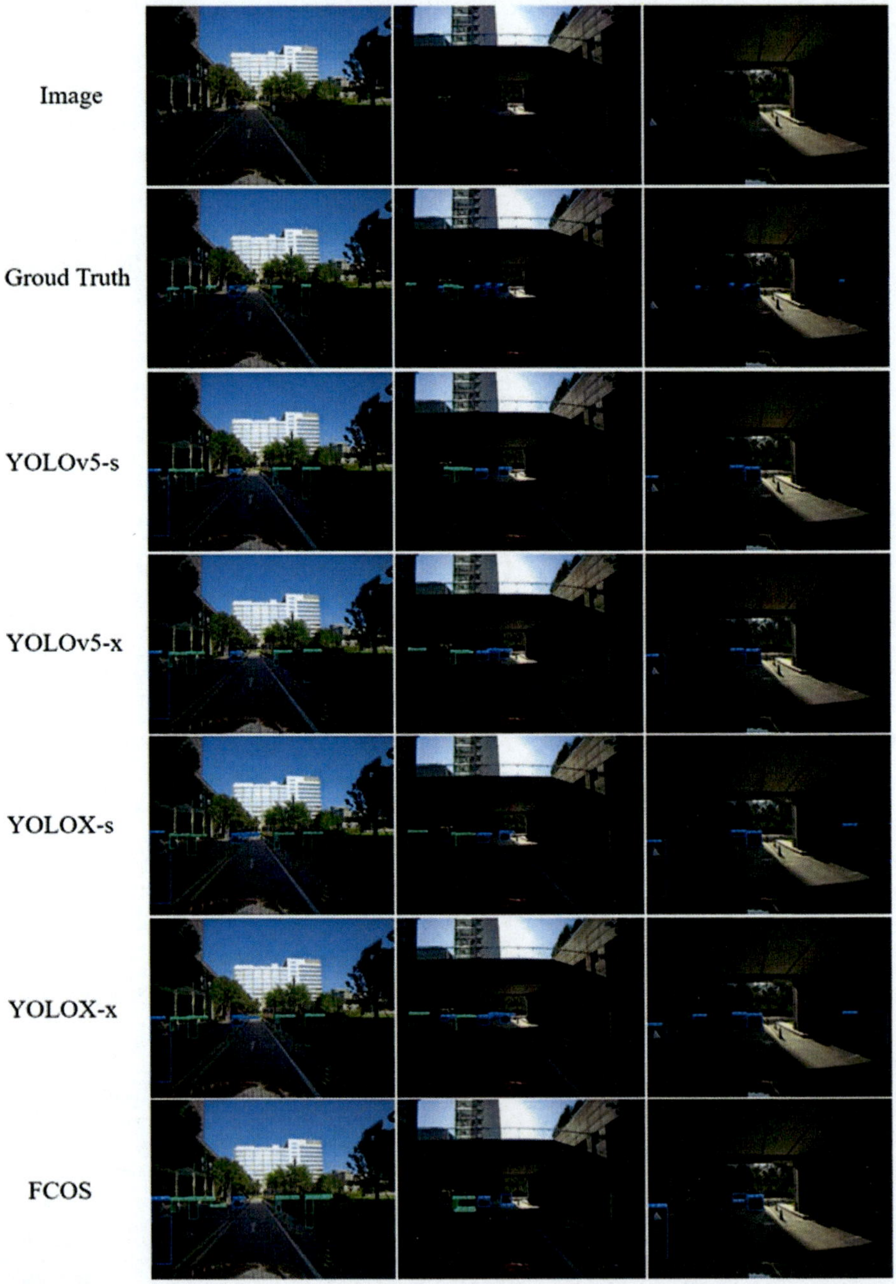

Fig. 7.9 2D object detection results of different methods

Table 7.7 Comparison of 3D object detection methods on OpenMPD and KITTI datasets. The "MOD" represents the moderate scene on KITTI

	OpenMPD	KITTI		
Model	mAP (%)	mAP(Mod) (%)	Speed(ms)	Device(GPU)
SECOND [44]	35.8	58.3	40	2.5 GHz
PointPillars [21]	32.5	60.8	**16**	GTX 1080 Ti
TANet	**40.3**	**62.0**	35	2.5 GHz

Bold values highlight the significant improvement in performance

Table 7.8 Comparison of 2D semantic segmentation methods on OpenMPD and COCO datasets. The speed, parameters, and flops are tested on a single RTX 3090

	OpenMPD	VOC			
Model	mIOU (%)	mIOU (%)	Speed (ms)	Parameters (M)	FLOPs(G)
DeepLabv3 [5]	**58.4**	85.7	76.9	39.76	512.87
DANet [10]	54.5	82.6	83.3	47.56	615.16
DUNet [40]	34.26	**88.1**	**58.8**	**29.38**	**391.79**

Bold values highlight the significant improvement in performance

7.5.2 Semantic Segmentation

For 2D semantic segmentation, we evaluate it using mean intersection over union (mIOU) followed by the Pascal VOC dataset [9]. The experiments aim at nine classes of interest (shown in Table 7.5) and one background class.

Similar to object detection, we use the pre-training weights of the VOC dataset to fine-tune DANet [10], DUNet [40], and DeepLabv3 [5]. The experimental results are shown in Table 7.8. From that, we found that the mIOU of the same method on OpenMPD is much lower than on the VOC dataset. In addition, the COCO segmentation has 20 classes, including indoor scenes, animals, pedestrians, vehicles, and so on. However, OpenMPD has only nine classes, which also reflects that OpenMPD has more complex scenes. Figure 7.10 shows the prediction of different methods and ground truth, from which we can see that the prediction results of each model in low light still have a certain gap compared with the ground truth, especially in the third column.

For 3D semantic segmentation, we also use mean intersection over union (mIOU) as an evaluating indicator, followed by the semantic KITTI dataset [2]. We train PolarNet [2], SalsaNext [8], and Cylinder3D [46] on OpenMPD and semantic KITTI, respectively. As shown in Table 7.9, the test results of the same method on OpenMPD are lower than those of semantic KITTI. On the one hand, we use 128-beam LiDAR, which means there are more dense point clouds, which increases the difficulty of fine-grained segmentation. On the other hand, the dataset we collected has more objects, and the scene is more complex.

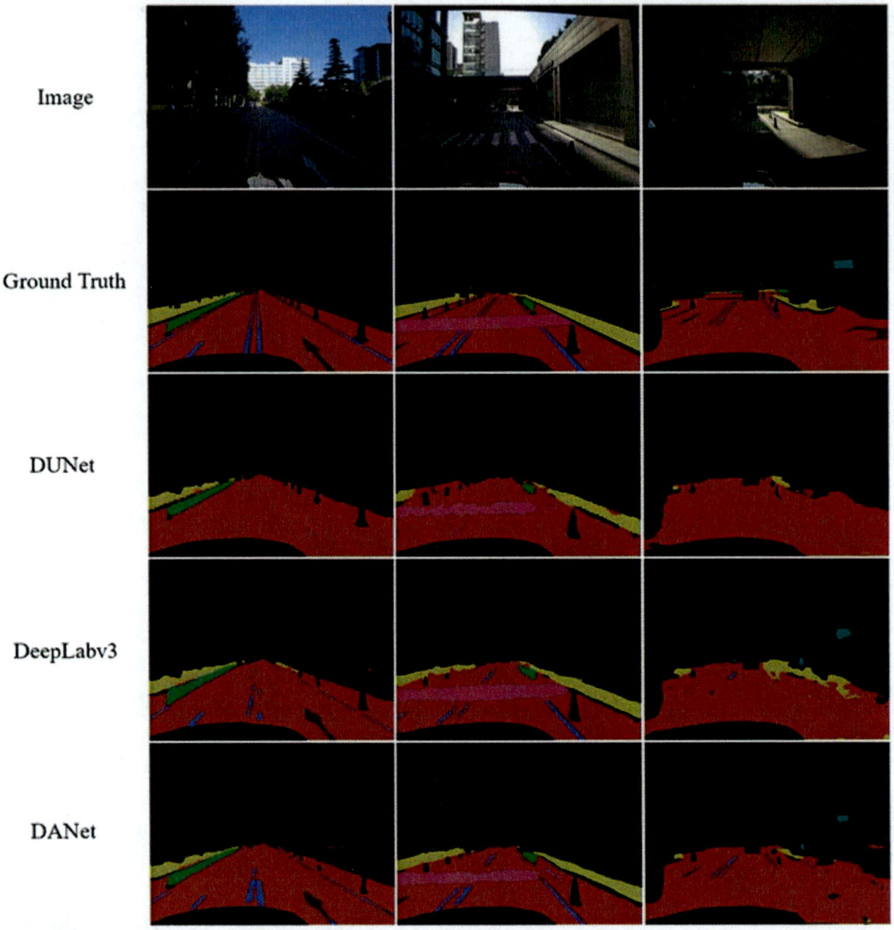

Fig. 7.10 2D semantic segmentation of different methods

Table 7.9 3D semantic segmentation of different methods

	OpenMPD	Semantic KITTI
Model	mIOU (%)	mIOU (%)
PolarNet [2]	57.6	71.0
SalsaNext [8]	60.2	72.2
Cylinder3D [46]	**61.8**	**76.1**

Bold values highlight the significant improvement in performance

7.6 Summary

In this chapter, we propose an autonomous driving dataset, OpenMPD. We provide a variety of annotations for 2D and 3D platforms, which can be used for object detection and semantic segmentation networks with multimodal (simultaneously using multiple sensors) and multi-task (simultaneously using detection and segmentation) fusion. We also provide a 360-degree shooting scene through six cameras, which can be combined with multi-view for object detection or segmentation tasks. OpenMPD plays a vital role in multi-task, multi-view, and multimodal tasks. Due to the continuous development of sensors, we use more advanced equipment compared with other datasets, which has a great impact on the quality of data, especially the 128-beam LiDAR. Considering the current autonomous driving sensing algorithms get good performance for simple scenes, we focus on complex scenes. A large number of baseline test results show that our dataset has a complex climate environment and complex road conditions. In addition, we also analyze the characteristics of the dataset to facilitate understanding.

In the future, we will use this dataset to test the autonomous driving system in our laboratory and further promote its application.

References

1. Alismail, H., Baker, L.D., Browning, B.: Continuous trajectory estimation for 3d slam from actuated lidar. In: IEEE International Conference on Robotics and Automation (2014)
2. Behley, J., Garbade, M., Milioto, A., Quenzel, J., Behnke, S., Stachniss, C., Gall, J.: SemanticKITTI: a dataset for semantic scene understanding of LiDAR sequences. In: Proc. of the IEEE/CVF International Conf. on Computer Vision (ICCV) (2019)
3. Behley, J., Steinhage, V., Cremers, A.B.: Performance of histogram descriptors for the classification of 3d laser range data in urban environments. In: Proceedings - IEEE International Conference on Robotics and Automation, pp. 4391–4398 (2012)
4. Caesar, H., Bankiti, V., Lang, A.H., Vora, S., Liong, V.E., Xu, Q., Krishnan, A., Pan, Y., Baldan, G., Beijbom, O.: nuscenes: a multimodal dataset for autonomous driving. In: Proceedings of the IEEE/CVF Conference on Computer Vision and Pattern Recognition, pp. 11,621–11,631 (2020)
5. Chen, L.C., Papandreou, G., Schroff, F., Adam, H.: Rethinking atrous convolution for semantic image segmentation. Preprint (2017). arXiv:1706.05587
6. Chen, X., Kundu, K., Zhang, Z., Ma, H., Urtasun, R.: Monocular 3d object detection for autonomous driving. In: IEEE Conference on Computer Vision and Pattern Recognition (2016)
7. Cordts, M., Omran, M., Ramos, S., Rehfeld, T., Schiele, B.: The cityscapes dataset for semantic urban scene understanding. In: 2016 IEEE Conference on Computer Vision and Pattern Recognition (CVPR) (2016)
8. Cortinhal, T., Tzelepis, G., Aksoy, E.E.: Salsanext: fast, uncertainty-aware semantic segmentation of lidar point clouds. In: International Symposium on Visual Computing, pp. 207–222. Springer (2020)
9. Everingham, M., Van Gool, L., Williams, C.K., Winn, J., Zisserman, A.: The pascal visual object classes (voc) challenge. Int. J. Comput. Vis. **88**(2), 303–338 (2010)

10. Fu, J., Liu, J., Tian, H., Li, Y., Bao, Y., Fang, Z., Lu, H.: Dual attention network for scene segmentation. In: Proceedings of the IEEE/CVF Conference on Computer Vision and Pattern Recognition, pp. 3146–3154 (2019)
11. Ge, Z., Liu, S., Wang, F., Li, Z., Sun, J.: Yolox: exceeding yolo series in 2021. Preprint (2021). arXiv:2107.08430
12. Geiger, A., Lenz, P., Stiller, C., Urtasun, R.: Vision meets robotics: the kitti dataset. Int. J. Robot. Res. **32**, 1231–1237 (2013)
13. Geyer, J., Kassahun, Y., Mahmudi, M., Ricou, X., Durgesh, R., Chung, A.S., Hauswald, L., Pham, V.H., Mühlegg, M., Dorn, S., et al.: A2d2: Audi autonomous driving dataset. Preprint (2020). arXiv:2004.06320
14. Hackel, T., Savinov, N., Ladicky, L., Wegner, J.D., Schindler, K., Pollefeys, M.: Semantic3d.net: a new large-scale point cloud classification benchmark. In: ISPRS Annals of Photogrammetry, Remote Sensing and Spatial Information Sciences (2017)
15. He, K., Gkioxari, G., Dollár, P., Girshick, R.: Mask R-CNN. In: IEEE (2017)
16. Hess, W., Kohler, D., Rapp, H., Andor, D.: Real-time loop closure in 2d lidar slam. In: 2016 IEEE International Conference on Robotics and Automation (ICRA) (2016)
17. Huang, X., Cheng, X., Geng, Q., Cao, B., Zhou, D., Wang, P., Lin, Y., Yang, R.: The apolloscape dataset for autonomous driving. In: 2018 IEEE/CVF Conference on Computer Vision and Pattern Recognition Workshops (CVPRW) pp. 1067–10,676 (2018)
18. Hwang, S., Park, J., Kim, N., Choi, Y., So Kweon, I.: Multispectral pedestrian detection: benchmark dataset and baseline. In: Proceedings of the IEEE Conference on Computer Vision and Pattern Recognition, pp. 1037–1045 (2015)
19. Jocher, G.: Yolov5 (2020). https://github.com/ultralytics/yolov5
20. Kohlbrecher, S., Stryk, O.V., Meyer, J., Klingauf, U.: A flexible and scalable slam system with full 3d motion estimation. In: IEEE International Symposium on Safety (2011)
21. Lang, A.H., Vora, S., Caesar, H., Zhou, L., Yang, J., Beijbom, O.: Pointpillars: fast encoders for object detection from point clouds. In: Proceedings of the IEEE/CVF Conference on Computer Vision and Pattern Recognition, pp. 12,697–12,705 (2019)
22. Leibe, B., Schiele, B.: Analyzing appearance and contour based methods for object categorization. In: Proceedings of the Computer Vision and Pattern Recognition, 2003. IEEE Computer Society (2003)
23. Lin, T.Y., Maire, M., Belongie, S., Hays, J., Zitnick, C.L.: Microsoft coco: common objects in context. In: European Conference on Computer Vision (2014)
24. López-Sastre, R.: Towards semantic and effective visual codebooks (2011). https://www.lap-publishing.com/
25. Moreels, P., Perona, P.: Evaluation of features detectors and descriptors based on 3d objects. Int. J. Comput. Vis. **73**(3), 263–284 (2007)
26. Mousavian, A., Anguelov, D., Flynn, J., Kosecka, J.: 3d bounding box estimation using deep learning and geometry. In: 2017 IEEE Conference on Computer Vision and Pattern Recognition (CVPR) (2017)
27. Munoz, D., Bagnell, J.A., Vandapel, N., Hebert, M.: Contextual classification with functional max-margin Markov networks. In: 2009 IEEE Conference on Computer Vision and Pattern Recognition (2009)
28. Neuhold, G., Ollmann, T., Bulo, S.R., Kontschieder, P.: The mapillary vistas dataset for semantic understanding of street scenes. In: IEEE International Conference on Computer Vision (2017)
29. Ozuysal, M., Lepetit, V., Fua, P.: Pose estimation for category specific multiview object localization. In: IEEE Conference on Computer Vision and Pattern Recognition (2009)
30. Pan, X., Shi, J., Luo, P., Wang, X., Tang, X.: Spatial as deep: Spatial cnn for traffic scene understanding. In: Thirty-Second AAAI Conference on Artificial Intelligence (2018)
31. Patil, A., Malla, S., Gang, H., Chen, Y.T.: The h3d dataset for full-surround 3d multi-object detection and tracking in crowded urban scenes. In: 2019 International Conference on Robotics and Automation (ICRA), pp. 9552–9557. IEEE (2019)

32. Pei Sun Henrik Kretzschmar, X.D.: Scalability in perception for autonomous driving: Waymo open dataset. In: Conference on Computer Vision and Pattern Recognition (CVPR) (2019)
33. Redmon, J., Farhadi, A.: Yolov3: an incremental improvement. arXiv e-prints (2018)
34. Ren, S., He, K., Girshick, R., Sun, J.: Faster R-CNN: towards real-time object detection with region proposal networks. IEEE Trans. Pattern Anal. Mach. Intell. **39**(6), 1137–1149 (2017)
35. Roddick, T., Kendall, A., Cipolla, R.: Orthographic feature transform for monocular 3d object detection. Preprint (2018). arXiv:1811.08188
36. Savarese, S., Li, F.F.: 3d generic object categorization, localization and pose estimation. In: IEEE 11th International Conference on Computer Vision, ICCV 2007, Rio de Janeiro, Brazil, October 14–20 (2007)
37. Steder, B., Ruhnke, M., Grzonka, S., Burgard, W.: Place recognition in 3d scans using a combination of bag of words and point feature based relative pose estimation. In: 2011 IEEE/RSJ International Conference on Intelligent Robots and Systems (2011)
38. Tan, M., Pang, R., Le, Q.V.: Efficientdet: scalable and efficient object detection. In: 2020 IEEE/CVF Conference on Computer Vision and Pattern Recognition (CVPR) (2020)
39. Thomas, A., Ferrar, V., Leibe, B., Tuytelaars, T., Schiel, B., Van Gool, L.: Towards multi-view object class detection. In: 2006 IEEE Computer Society Conference on Computer Vision and Pattern Recognition (CVPR'06), vol. 2, pp. 1589–1596. IEEE (2006)
40. Tian, Z., He, T., Shen, C., Yan, Y.: Decoders matter for semantic segmentation: Data-dependent decoding enables flexible feature aggregation. In: Proceedings of the IEEE/CVF Conference on Computer Vision and Pattern Recognition, pp. 3126–3135 (2019)
41. Tian, Z., Shen, C., Chen, H., He, T.: Fcos: fully convolutional one-stage object detection. In: Proceedings of the IEEE/CVF International Conference on Computer Vision, pp. 9627–9636 (2019)
42. Xiaozhi, C., Kaustav, K., Yukun, Z., Huimin, M., Sanja, F.: 3d object proposals using stereo imagery for accurate object class detection. IEEE Trans. Pattern Anal. Mach. Intell. **PP**(99), 1–1 (2017)
43. Xu, B., Chen, Z.: Multi-level fusion based 3d object detection from monocular images. In: 2018 IEEE/CVF Conference on Computer Vision and Pattern Recognition (CVPR) (2018)
44. Yan, Y., Mao, Y., Li, B.: Second: sparsely embedded convolutional detection. Sensors **18**(10), 3337 (2018)
45. Zhang, R., Candra, S.A., Kai, V., Zakhor, A.: Sensor fusion for semantic segmentation of urban scenes. In: IEEE International Conference on Robotics and Automation (2015)
46. Zhu, X., Zhou, H., Wang, T., Hong, F., Ma, Y., Li, W., Li, H., Lin, D.: Cylindrical and asymmetrical 3d convolution networks for lidar segmentation. In: Proceedings of the IEEE/CVF Conference on Computer Vision and Pattern Recognition, pp. 9939–9948 (2021)

Chapter 8
Vehicle-Road Multi-View Interactive Data Fusion

Abstract Aiming at solving the problem of target occlusion in crowded scenes, the aggregation of multiple camera views can outperform a single view. Some researchers project the feature map or RGB image getting from the feature extraction network to the ground plane by perspective transformation and then use the spatial aggregation without anchor frame to aggregate the feature maps of multiple views. However, the projected feature maps are limited by a lack of semantic information. In order to obtain feature maps with rich semantic information, we will introduce a new multi-view detection method MVT, which effectively extracts features from multiple perspectives. To improve TP and the overall robustness of the model, we will also introduce a method for predicting the positioning uncertainty of bounding box reliability.

This chapter begins by describing the background of detection methods from three perspectives—single-perspective method, multi-view method, and transformers method—before delving into the methodology and experimental results and finally providing a summary of the methods used in the experiments.

8.1 Introduction

MVDet [7], the best performing multi-view detection method at the moment, extracts features from multiple views using a common convolutional neural network. The outcome of feature extraction has a significant impact on subsequent detection. The multi-view feature fusion will be prone to distortion if the feature information of each view is not sufficiently extracted. Furthermore, the pattern of distortion varies depending on camera positions or projection processes. The inadequacy of semantic features may also have a limiting effect on the detection system as a whole (Fig. 8.1).

Single-Perspective Method The majority of target detection methods are based on a single camera and rely heavily on deep learning methods. Traditional target

X. Zhang et al., *Multi-sensor Fusion for Autonomous Driving*,
https://doi.org/10.1007/978-981-99-3280-1_8

Fig. 8.1 The multi-view detection system observes the position of each target in the form of a bird's-eye view

detection algorithms include one-stage and two-stage algorithms. When compared to second-stage algorithms, most first-stage algorithms have lower accuracy but faster speeds. Many common algorithms are based on anchor frames, but some researchers employ anchorless methods CornerNet [9] and FCOS [13]. NMS is also studied by some researchers, such as GrooMeD-NMS [8] and MDS-NMS [12]. However, in intensive detection or intersections, these methods will encounter occlusion issues, as well as missed detection, false detection, and low detection accuracy. Our research aims to find effective solutions to these problems.

Multi-View Method Occlusion can be effectively solved using multi-view applications. With a single camera, object overlap or occlusion occurs, whereas multiple cameras can spatially aggregate their respective images. Many researchers employ spatial aggregation methods when processing image data. MVDet, for example, performs perspective transformation on the feature map produced by CNN and then spatially aggregates multiple views to achieve good results. To obtain world features, we project feature maps and splice multi-view feature maps using the perspective transformation method.

Transformers Method Transformers are currently being used by researchers to replace the feature extraction structure of convolutional neural networks [2, 11, 14, 15, 17], specifically the encoder-decoder structure. The benefits of the inspection process are simplified because anchor frames and other artificial setting factors are no longer required, allowing it to shine in computer vision. Our method is based on the deformable transformers concept, which combines deformable convolutional sparse spatial sampling and transformers. The benefits of relational modeling capabilities improve multi-view image feature extraction and lay the groundwork for later detection and use.

This chapter employs a non-anchor frame optimization feature extraction method to improve multi-view detection and positioning accuracy. Using the encoder-decoder structure's effective attention mechanism, we replace ordinary convolutions with deformable transformers and effectively extract features from multiple views. During the feature extraction process, we embed corner pooling, improve the semantic information of the target corner position, and analyze the upper left and lower right corner positions. We also took into account the loss of positioning uncertainty. By incorporating predicted positioning uncertainty into the detection process, the FP is effectively reduced, while the TP and positioning accuracy are improved.

8.2 Methodology

This chapter will describe the structure of our method, as illustrated in Fig. 8.2. To begin, we propose replacing ordinary convolutional neural networks with deformable transformers to extract features from multi-view images in blocks. When compared to CNN, it simplifies the settings of a large number of hyperparameters in the detection process. Furthermore, the incorporation of deformable attention makes the feature extraction part more concerned. Semantic information surrounding key points improves the ability of multi-view representation. Then we borrowed the idea of a residual network and proposed the residual corner pooling module, which replaced the maximum pooling layer in ResNet50 and later enhanced the feature map's corner semantic information, laying the groundwork for subsequent multi-view detection. Taking into account the impact of loss due to bounding box, we add positioning uncertainty to the loss, which can reduce FP, increase TP, improve the accuracy of multi-view pedestrian positioning, and increase the model's robustness. The multi-view feature maps are projected and aggregated into a new bird's-eye view.

Next, the method's composition will be introduced. The perspective transformation method in multi-view detection will be introduced in Sect. 8.1. Section 8.2 will discuss the applicability of the deformable transformers method in multi-view detection. In Sect. 8.3, we will go over our residual corners in detail, as well as the significance of the pooling module. The role of positioning uncertainty loss in the training model will be introduced in Sect. 8.4.

We continue to use the perspective transformation method of MVDET [9] to project the feature map. First, we convert each pixel in the feature map from three-dimensional coordinates (x, y, z) to two-dimensional coordinates (u, v). The perspective transformation Eq. 8.1 briefly introduces the perspective transformation method. s is the scale factor, P_θ is the perspective transformation matrix, A is the 3×3 internal parameter matrix, $[R \mid t]$ is the 3×4 external parameter matrix, R is the rotation factor, and t is the translation parameter.

$$s\begin{pmatrix}u\\v\\1\end{pmatrix}=P_\theta\begin{pmatrix}x\\y\\z\\1\end{pmatrix}=A[R|t]\begin{pmatrix}x\\y\\z\\1\end{pmatrix}=\begin{bmatrix}\theta_{11}\theta_{12}\theta_{13}\theta_{14}\\\theta_{21}\theta_{22}\theta_{23}\theta_{24}\\\theta_{31}\theta_{32}\theta_{33}\theta_{34}\end{bmatrix}\begin{pmatrix}x\\y\\z\\1\end{pmatrix} \tag{8.1}$$

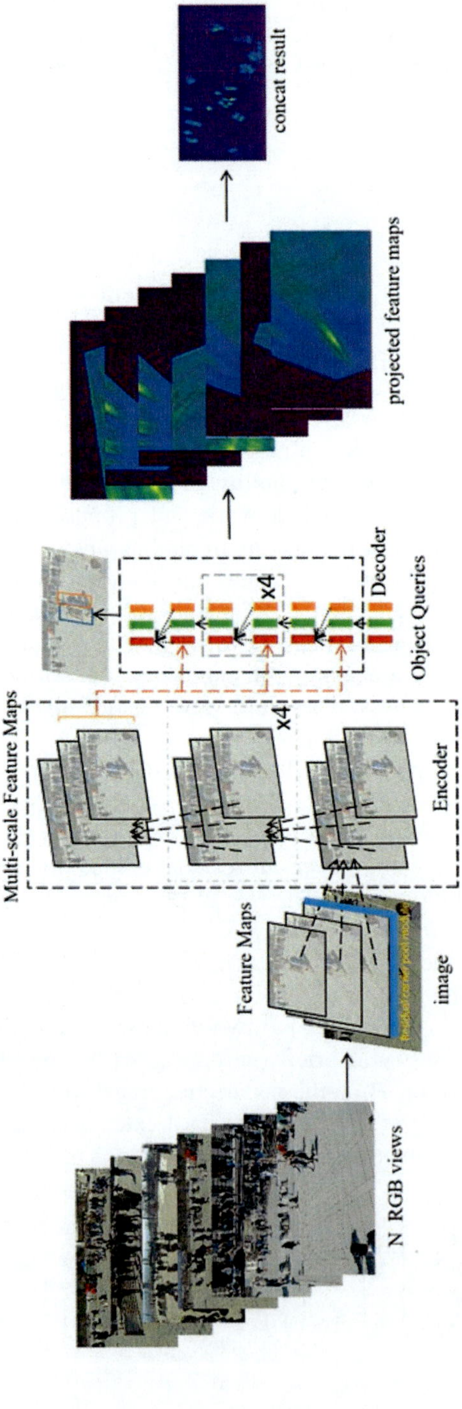

Fig. 8.2 The frame structure of our method

To obtain [H,W] size, (NxC+2) channel ground plane feature maps, we project C channel feature maps into N cameras. Given the issue of ground plane neighbors, we maintain a larger ground plane receptive field by using a large convolution kernel, which effectively aggregates spatial information.

Transformers are currently divided into two types: those made entirely of transformers and those made of convolutional neural networks and transformers. The proposal of DETR [17] reduces a large number of parameters, greatly simplifying the detection process and the setting of the anchor frame. However, it suffers from slow convergence speed and poor detection performance of small targets. Deformable DETR [10] solves the problem of slow DETR convergence by combining the benefits of deformable convolution [17] with sparse spatial sampling and transformers' relational modeling capabilities. We only use the previous convolutional neural network and encoder-decoder structure, leaving out the prediction module. The encoder-decoder structure is embedded with a deformable attention module in deformable transformers. Regardless of the spatial size of the feature map, the deformable attention module only focuses on a small set of key sampling points around the reference point. Convergence and feature spatial resolution issues can be alleviated by assigning only a small fixed number of key points to each query.

We replace the multi-scale deformable attention module with the transform attention module that deals with feature mapping in DETR. The encoder's input and output are multi-scale feature maps with the same resolution. Multi-scale feature maps are generated in the encoder from feature maps in the ResNet network's C3–C5 stages. The feature maps of the C3–C5 stages are subjected to a 1×1 convolution operation with a step size of 1 to generate a feature map of the same size and 256 channels, and then C5 is subjected to a 3×3 convolution with a step size of 2. The sampling convolution operation, as C6, generates a half-size multi-scale feature map. Because the proposed multi-scale deformed attention can exchange information between multi-scale feature maps, these feature maps used the multi-scale concept rather than the top-down structure of FPN [10]. As a result, FPN is not employed in this chapter. The encoder produces a multi-scale feature map with the same resolution as the input when the multi-scale deformable attention module is applied. All of the key and query elements are pixels from the multi-scale feature map. The reference point for each query pixel is itself. In order to determine the feature level of each query pixel, we add a scale-level embedding to the feature representation in addition to the position embedding. In contrast to fixed coding position embedding, scale-level embedding is randomized and trained alongside the network.

The decoder includes cross-attention and self-attention modules. Object queries are the query elements of these two types of attention modules. The object query in the cross-attention module extracts features from the feature map, where the key element is the encoder's output feature map. In the self-attention module, object queries interact with one another, with the object query being the most important. Because our proposed deformable attention module is intended to process convolutional feature maps as a key component, we only replace each cross-attention module with a multi-scale deformable attention module, leaving the self-attention module alone.

Corner pooling can improve the semantic information of corner points, effectively focusing the target's corner points. Corner pooling is typically divided into two types: top-left corner pooling and bottom-right corner pooling. Formula 8.2 demonstrates the specific principle of corner pooling.

$$
\begin{aligned}
t_{ij} &= \begin{cases} \max\left(f_{t_i}, t_{(i+1)j}\right) & \text{if } i < H \\ f_{t_{\text{fj}}} & \text{otherwise} \end{cases} \\
l_{ij} &= \begin{cases} \max\left(f_{l_{ij}}, l_{j(i+1)}\right) & \text{if } j < H \\ f_{l_{iv}} & \text{otherwise} \end{cases}
\end{aligned}
\tag{8.2}
$$

Assuming that the size of the feature map is $H \times W$, and each pixel in the figure is t_{ij}, the pixel points are compared along the top and left directions. If it is less than the pixel in this direction, the pixel size is inherent; if it is greater, pixels in this direction cover pixels in this direction. As a result, there is an increasing sequence in the top and left directions, and the small pixels that existed previously will be covered by nearby ones, significantly improving the semantic information.

CornerNet [13] inspired us to optimize top-left corner pooling, and the optimization structure for top-left corner pooling is shown in Fig. 8.3. The feature maps are grouped into two directions: top and left. The two feature maps are combined to produce two feature maps with enhanced semantic information. Finally, a map of pooling features in the top-left corner is obtained. The characteristics of the two feature maps are combined in this feature map. The top corner pooling feature map increases the pixel value in the top direction gradually. Similarly, the left corner pooling feature map gradually increases the number of pixel points to the left. When the two feature maps are fused, the semantic information in the upper left corner is stronger than the semantic information in the two sub-corner pooled feature maps.

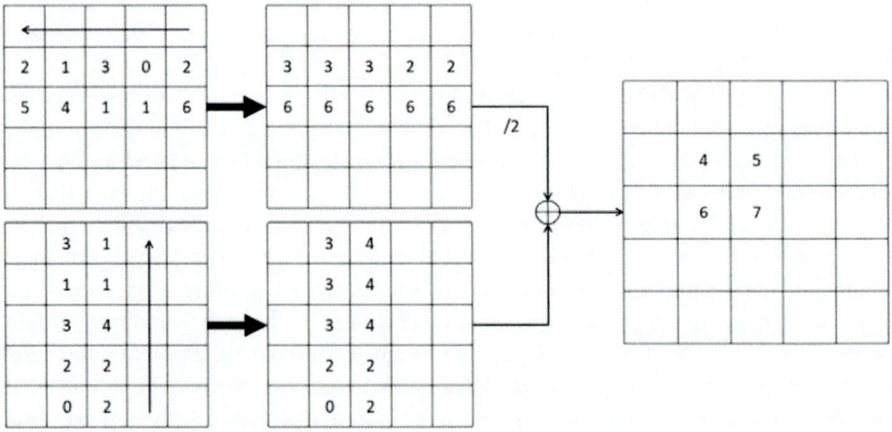

Fig. 8.3 Top-left corner pooling structure

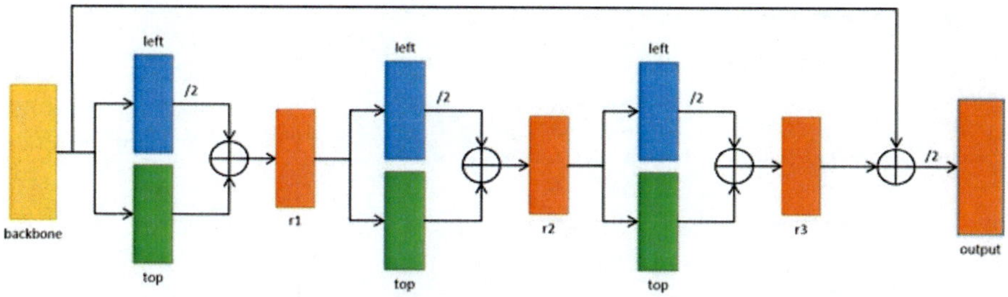

Fig. 8.4 Residual corner point pooling module. The top-left corner point pooling is used for three consecutive layers, and the initial convolution feature map is jump-connected with the resulting feature map r3 to obtain the latest feature map output

The distinction between the upper left corner position and the other pixels will be more visible. The semantic information in the upper left corner will be enhanced.

As shown in Fig. 8.4, we propose a new three-layer jump connection corner pooling module. ResNet50 serves as the foundation. Because the initial feature map location is rich in semantic information, we perform corner pooling operations on it. First, we use this module to replace ResNet50's first maximum pooling layer. Then, for three consecutive layers, the top-left corner point pooling is used, and the initial convolution feature map is jump-connected with the resulting feature map r3 to obtain the most recent feature map output. According to studies, each additional layer of corner point pooling strengthens the semantic information of the upper left corner of the final feature map, but excessive layers can easily result in the loss of local location semantic information. Finally, the final feature map is sent to the following convolutional layer for feature extraction.

We draw on the idea of positioning uncertainty of Gaussian YOLOv3 [16] and consider it along with target score and category score [4]. The traditional detection model is reduced to four dimensions, namely, the prediction frame's center point and aspect. The detection model of the integration uncertainty has eight dimensions, which correspond to the predictive frame's central coordinates and long width, as well as the uncertainty of the corresponding prediction box. The mean of the Gaussian distribution is the target box's position, and the corresponding uncertainty is variance. In order to predict the uncertainty of bounding box, each bounding box coordinate model in the predictive feature is mean μ and variance Σ, as shown in Fig. 8.5.

$$
\begin{aligned}
&\mu_{t_x} = \sigma\left(\hat{\mu}_{t_x}\right), \mu_{t_y} = \sigma\left(\hat{\mu}_{t_y}\right), \mu_{t_w} = \hat{\mu}_{t_w}, \mu_{t_h} = \hat{\mu}_{t_h} \\
&\Sigma_{t_x} = \sigma\left(\hat{\Sigma}_{t_x}\right), \Sigma_{t_v} = \sigma\left(\hat{\Sigma}_{t_y}\right) \\
&\Sigma_{t_w} = \sigma\left(\hat{\Sigma}_{t_w}\right), \Sigma_{t_h} = \sigma\left(\hat{\Sigma}_{t_h}\right) \\
&\sigma(x) = \frac{1}{(1+\exp(-x))}.
\end{aligned}
\tag{8.3}
$$

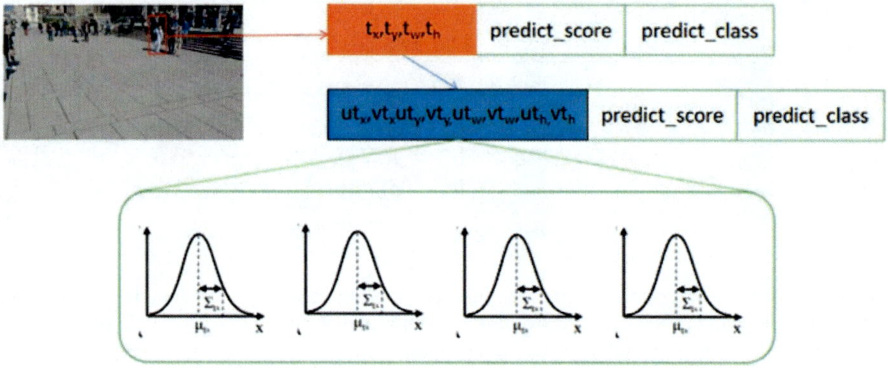

Fig. 8.5 Components in the prediction box of positioning uncertainty method

The bounding box's predictive coordinates are the average values of each coordinate in the detection layer, with each variance representing the uncertainty of each coordinate. Because Eq. 8.3 represents the central coordinates of the grid's bounding box, we use the sigmoid function to process values between 0 and 1. The variance of each coordinate is also converted into a value between 0 and 1. We can calculate the bounding box uncertainty of each detected object in the image using the sigmoid function. Because the uncertainty does not apply to the entire image, it can be applied to each detection result. MVDet [7] only takes into account the target score and class score during object detection. It can't consider the bounding box score during the detection process because the score information for the bounding box coordinates is unknown. Positioning uncertainty is represented by the bounding box score. During the detection process, positioning uncertainty, as well as objective and category scores, can be taken into account. The positioning uncertainty is used in multi-view detection to filter the bounding box with high uncertainty in the prediction result during the detection process. In this manner, a prediction with high confidence in objectness, class, and bounding box is finally chosen. Uncertainty loss can reduce FP while increasing TP, resulting in improved detection accuracy. The standard formula for detecting positioning uncertainty is as follows:

$$S = \sigma(\text{ Object }) \times \sigma(\text{ Class }_i) \times \left(1 - \text{ Uncertainy }_{\text{aver}}\right) \tag{8.4}$$

In (5), S represents the final detection standard value, $\sigma(\text{ Object })$ represents the objectness score, and $\sigma(\text{ Class }_i)$ represents the $i$class score. Furthermore, Uncertaintyaver represents the average degree of uncertainty in predicted bounding box coordinates. Positioning uncertainty, like objectivity and class score, has a value between 0 and 1. Higher positioning uncertainty decreases confidence in the predicted bounding box and decreases the IoU between the predicted bounding box and the real box. Lower positioning uncertainty results in more confidence in the predicted bounding box and a smaller IoU between the predicted and real boxes. The positioning uncertainty effectively represents the predictability of the bounding box.

8.3 Experiment

Datasets MVDet [7] references the Wildtrack dataset [9] and the MultiviewX dataset. The characteristics of the two datasets are also introduced in detail, and we will continue to investigate them. Table 8.1 compares data from the Wildtrack and MultiviewX datasets. Figure 8.6 shows an example of the Wildtrack and MultiviewX datasets.

Metrics To evaluate the algorithm, we use precision, recall, MODA, and MODP. MODP is used to assess positioning accuracy, whereas MODA is used to explain false positives and false negatives. MODA is used as the primary performance indicator in this experiment because it explains both false positives and false negatives. To see if it is truly positive, we set the threshold to 0.5.

Implementation Details For model training, we use Ubuntu 16.04 operating system, PyTorch 1.7, and two graphics memory 3090 with a total of 32G RTX. In the training phase, considering the large number of images, we resize the original image 1080×1920 to 675×1200 and batch size to 1, which is convenient for model training. We use the SGD optimizer with a momentum of 0.5 and L2 normalization to 5×10^{-4}, and the single-view loss weight α is set to 1. We employ a single-cycle learning rate scheduler with a maximum learning rate of 0.1 and a maximum number of epochs of 50; we also employ deformable transformers as the skeleton of our method, with the number of encoders in transformers set to six.

Experiments are carried out on the Wildtrack dataset and the MultiviewX dataset. We compare RCNN and clustering [6], POM-CNN [3], DeepMCD [1], Deep-Occlusion [5], and MVDet [7] with our method. Among them, MVDet is the experiment's baseline, and we will compare them to it. MVT (deformable transformers) is a method that only uses deformable transformers, whereas MVT (w/o

Table 8.1 Comparison between Wildtrack dataset and MultiviewX dataset

	# camera	Resolution	Frames	Area
Wildtrack	7	1080×1920	400	$12\times36\,m^2$
MultiviewX	6	1080×1920	400	$16\times25\,m^2$

(a) Wildtrack (b) MultiviewX

Fig. 8.6 Sample pictures in the Wildtrack dataset and MultiviewX dataset residual corner pooling module) does not use CornerBlock but does use deformable transformers and positioning uncertainty. Positioning uncertainty is not included in MVT (w/o localization uncertainty), but it does include residual corner pooling module and deformable transformers. Our MVT method incorporates deformable transformers, a residual corner pooling module, and localization uncertainty. The experimental comparison results of our method and other classic methods on the Wildtrack and MultiviewX datasets are shown in Table 8.2.

The Influence of Deformable Transformers Deformable transformers are used to implement the feature extraction function in Fig. 8.7. The embedding of deformable attention directs more attention to a small group of key sampling points surrounding the reference point, improving the ability to model relationships. According to

Table 8.2 Some experimental comparison results on the Wildtrack and MultiviewX datasets

	Wildtrack				MultiviewX			
Method	MODA	MODP	Precision	Recall	MODA	MODP	Precision	Recall
RCNN and clustering	11.3	18.4	68.0	43.0	18.7	46.4	63.5	43.9
POM-CNN	23.2	30.5	75.0	55.0	–	–	–	–
DeepMCD	67.8	64.2	85.0	82.0	70.0	73	85.7	83.3
Deep-Occlusion	74.1	53.8	95.0	80.0	75.2	54.7	97.8	80.2
MVDet [3]	88.2	**75.7**	94.7	93.6	83.9	79.6	96.8	86.7
MVT (deformable transformers)	81.9	72.7	93.8	87.7	68.7	75.3	92.4	74.8
MVT (w/o residual corner pooling module)	88.8	74.6	95.6	93.1	87.6	80.9	**97.2**	90.2
MVT (w/o localization uncertainty)	80.7	71.7	92.6	87.7	68.1	74.3	91.7	74.9
MVT	**89.9**	74.9	**95.7**	**94.1**	**88.5**	**81.5**	96.7	**91.6**

Bold values highlight the significant improvement in performance

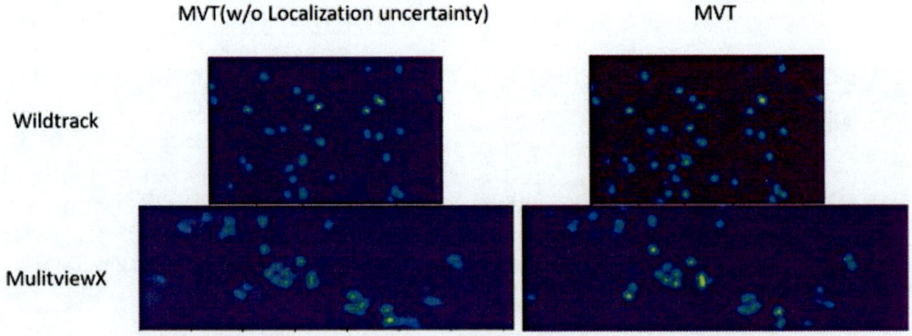

Fig. 8.7 MVT (w/o localization uncertainty) and MVT method

the experimental results in Table 8.2, MVT (deformable transformers) achieved 81.9% MODA and 87.7% recall on the Wildtrack dataset and 68.7% MODA and 74.8% recall on the MultiviewX dataset. MVT's performance on the two datasets is significantly lower than that of MVDet. Deformable transformers have a better effect in single-view detection, but in multi-view detection, multiple pictures may suffer feature loss due to the encoder-decoder, affecting subsequent multi-view aggregation. We add MVT (w/o residual corner pooling module) and MVT (w/o localization uncertainty) variant ablation experiments to the MVT (deformable transformers) method. Experiments show that the two variants perform differently in terms of accuracy than MVT (deformable transformers). Because the loss calculation was not fully considered, the accuracy of MVT (w/o localization uncertainty) in the two datasets was reduced when compared to MVT (deformable transformers). MVT (w/o localization uncertainty) accuracy has significantly improved in both datasets. It can be seen that ordinary convolutional neural networks can be replaced with deformable transformers; overall, deformable transformers, as the backbone network of our method, greatly improve the model's characterization ability and robustness and also lay the groundwork for later feature processing.

The Influence of the Number of Different Corner Pooling Layers in the Residual Corner Pooling Module To improve the semantic information of the upper left corner, we use the jumping corner pooling module to replace the maximum pooling layer in the basic CNN, as shown in Fig. 8.7. According to the findings, the number of different corner pooling layers has a significant impact on overall detection accuracy. Table 8.3 shows how the number of corner pooling layers in the corner module affects detection performance. We can see that as the number of corner pooling layers increases, so does the accuracy achieved by MODA. Furthermore, we gradually increase the number of corner pooling layers in the residual corner module, and the accuracy gradually improves, however, after adding three corner pooling layers, MODA achieved 89.9% accuracy in the Wildtrack dataset, and that of Recall was as high as 94.1%. In the MultiviewX dataset, MODA achieves 88.5% accuracy, and recall is as high as 91.6%. The effect, however, will be diminished if the corner pooling layer is added later. As shown in Fig. 8.4, the C1 feature map is merged with the feature map after three layers of corner point pooling, with the result being more inclined to the corner point position with each layer of corner

Table 8.3 The influence of the number of corner pooling layers in the residual corner module on the detection performance

	Wildtrack				MultiviewX			
#corner_layers	MODA	MODP	Precision	Recall	MODA	MODP	Precision	Recall
1	88.4	76	96.9	91.4	87.2	81	97.6	89.4
2	88.6	73.4	95.8	92.6	87.5	80.7	95.8	91.5
3	89.9	74.9	95.7	94.1	88.5	81.5	96.7	91.6
4	89.5	74.5	95.5	93.9	87.7	80.4	96.5	91.0

point pooling. The end result is the same as the weight. When you add it to the C1 feature map, you'll get a new corner feature map. The new corner feature map is naturally used in the subsequent feature extraction, which has a better overall improvement in the multi-view detection effect after the experiment. Corner pooling is equivalent to a pixel-level semantic information weighting operation, and it is also a way to enhance pixel semantic information to improve the effect of feature extraction, but too many layers can easily result in the loss of target semantic information. We discovered that when deformable transformers and residual corner pooling modules were included in the MVT (w/o localization uncertainty) variant experiment, the accuracy dropped. MVT (w/o localization uncertainty) achieved 80.7% accuracy of MODA in Wildtrack dataset and 87.7% of recall. In MultiviewX dataset, MODA achieved 68.1% accuracy and recall of 74.9%. The residual corner pooling module and localization uncertainty have different effects after analysis. The residual corner pooling module optimizes forward propagation, but it is not absolute, as demonstrated by the experimental results in Table 8.3. The accuracy obtained in the two datasets is close to MVT (w/o residual corner pooling module). It can be seen that, while the residual corner pooling module can optimize feature extraction across multiple views, its impact on the overall multi-view detection model is limited.

The Impact of Localization Uncertainty Because MVDet [7] does not take the bbox score into account during the detection process, the bbox score information is unknown. The positioning uncertainty [16] takes the bounding box loss of each detection object into account. In this chapter, we introduce positioning uncertainty as well as bbox loss and other classification losses. It also helps to improve our loss function. We can improve positioning accuracy, reduce FP, and increase TP by introducing positioning uncertainty. Recall has improved significantly. Table 8.2 shows that the method MVT in this chapter has positioning uncertainty. Among them, 89.9% of MODA and 94.1% of recall were obtained on the Wildtrack dataset; 88.5% of MODA and 91.6% of recall were obtained on the MultiviewX dataset. Although the MVT (w/o localization uncertainty) variant does not have positioning uncertainty, the accuracy obtained is significantly lower than that of the other variants. The experimental results show that localization uncertainty is crucial in this method. By comparing the experimental results of MVT (w/o localization uncertainty) with other variants, we discovered that the overall model's accuracy is very low. Figure 8.7 shows the MVT results map (without localization uncertainty and MVT). We thoroughly investigate the impact of localization uncertainty on detection and positioning and discover that it has a significant impact on the detection and positioning of a single-view image. Multiple perspectives superimposed has a superposition effect on positioning uncertainty as well. The goal of multi-view detection experiments is to show that the location has a direct influence on the experiment's results. As a result, we added positioning uncertainty to our algorithm model, optimized the model's overall loss function, and improved its robustness.

8.4 Summary

One-click annotation is a simple and effective 2D annotation method that we proposed. This method only requires the user to provide a click close to the center of the object. In the current mainstream semi-automatic annotation tools, IoU of annotation boxes performs best. Furthermore, our method naturally supports the interactive annotation of additional points for additional correction. Despite its simplicity, numerous experiments have demonstrated that our model generalizes well across datasets and fields, demonstrating its superiority as an annotation tool. We intend to investigate the domain of adaptive semi-automatic labeling in the future, and a single click can annotate all objects in the image that are in the same category as the clicked object. We believe that our research will stimulate new research ideas in the field of automatic labeling. In this chapter, we use a new method to improve the accuracy of multi-view detection. We replace ordinary convolutional networks with deformable transformers and use the spatial sparsity capability of deformable convolutions and the modeling relationship of transformers to effectively extract image features from multiple cameras. In order to make better use of the feature information from multiple cameras, we chose a new residual corner pooling module to enhance pixel-level semantic information and highlight the corner positions. Finally, we also used the loss of positioning uncertainty to supplement the loss of bounding box, which reduced FP, increased TP, improved positioning accuracy, and optimized model convergence. The accuracy of our method reached 88.5 and 89.9% in the MultiviewX dataset and Wildtrack dataset, respectively, and achieved good competitive results. We also hope to have further prospects.

References

1. Baqué, P., Fleuret, F., Fua, P.: Deep occlusion reasoning for multi-camera multi-target detection. In: Proceedings of the IEEE International Conference on Computer Vision, pp. 271–279 (2017)
2. Carion, N., Massa, F., Synnaeve, G., Usunier, N., Kirillov, A., Zagoruyko, S.: End-to-end object detection with transformers. In: European Conference on Computer Vision, pp. 213–229. Springer, Berlin (2020)
3. Chavdarova, T., Fleuret, F.: Deep multi-camera people detection. In: 2017 16th IEEE International Conference on Machine Learning and Applications (ICMLA), pp. 848–853. IEEE, Piscaataway (2017)
4. Chen, W., Xu, J., Zhao, X., Liu, Y., Yang, J.: Separated sonar localization system for indoor robot navigation. IEEE Trans. Ind. Electron. **68**(7), 6042–6052 (2020)
5. Fang, Y., Ding, G., Wen, W., Yuan, F., Yang, Y., Fang, Z., Lin, W.: Salient object detection by spatiotemporal and semantic features in real-time video processing systems. IEEE Trans. Ind. Electron. **67**(11), 9893–9903 (2019)
6. Fleuret, F., Berclaz, J., Lengagne, R., Fua, P.: Multicamera people tracking with a probabilistic occupancy map. IEEE Trans. Pattern Analy. Mach. Intell. **30**(2), 267–282 (2007)

7. Hou, Y., Zheng, L., Gould, S.: Multiview detection with feature perspective transformation. In: European Conference on Computer Vision, pp. 1–18. Springer, Berlin (2020)
8. Kumar, A., Brazil, G., Liu, X.: GrooMeD-NMS: Grouped mathematically differentiable nms for monocular 3d object detection. In: Proceedings of the IEEE/CVF Conference on Computer Vision and Pattern Recognition, pp. 8973–8983 (2021)
9. Law, H., Deng, J.: Cornernet: Detecting objects as paired keypoints. In: Proceedings of the European Conference on Computer Vision (ECCV), pp. 734–750 (2018)
10. Lin, T.Y., Dollár, P., Girshick, R., He, K., Hariharan, B., Belongie, S.: Feature pyramid networks for object detection. In: Proceedings of the IEEE Conference on Computer Vision and Pattern Recognition, pp. 2117–2125 (2017)
11. Liu, Z., Lin, Y., Cao, Y., Hu, H., Wei, Y., Zhang, Z., Lin, S., Guo, B.: Swin transformer: Hierarchical vision transformer using shifted windows. In: Proceedings of the IEEE/CVF International Conference on Computer Vision, pp. 10012–10022 (2021)
12. Rodriguez-Blazquez, C., Schrag, A., Rizos, A., Chaudhuri, K.R., Martinez-Martin, P., Weintraub, D.: Prevalence of non-motor symptoms and non-motor fluctuations in parkinson's disease using the MDS-NMS. Movement Disorders Clini. Pract. **8**(2), 231–239 (2021)
13. Tian, Z., Shen, C., Chen, H., He, T.: FCOS: Fully convolutional one-stage object detection. In: Proceedings of the IEEE/CVF International Conference on Computer Vision, pp. 9627–9636 (2019)
14. Wu, H., Xiao, B., Codella, N., Liu, M., Dai, X., Yuan, L., Zhang, L.: CvT: Introducing convolutions to vision transformers. In: Proceedings of the IEEE/CVF International Conference on Computer Vision, pp. 22–31 (2021)
15. Wu, S., Wu, T., Lin, F., Tian, S., Guo, G.: Fully transformer networks for semantic image segmentation (2021). Preprint arXiv:2106.04108
16. Xu, Y., Liu, X., Liu, Y., Zhu, S.C.: Multi-view people tracking via hierarchical trajectory composition. In: Proceedings of the IEEE Conference on Computer Vision and Pattern Recognition, pp. 4256–4265 (2016)
17. Zhu, X., Su, W., Lu, L., Li, B., Wang, X., Dai, J.: Deformable DETR: Deformable transformers for end-to-end object detection (2020). Preprint arXiv:2010.04159

Chapter 9
Information Quality in Data Fusion

Abstract As an essential element of multi-sensor fusion, uncertainty provides theoretical support for making better decisions using multi-source data. This chapter begins by introducing the prerequisite knowledge. Following that, two types of uncertainty are introduced from the perspective of models and data, known as epistemic uncertainty and theoretical uncertainty. It is possible to deal with the contingencies for a given model or set of sensors on a vehicle by delving into these two types of uncertainty. This chapter proposes a novel multimodal fusion architecture from an information-theoretic perspective. The proposed model is presented from four aspects: baseline, uncertainty modeling, fusion step, and implementation. Furthermore, the experimental procedure, including data preprocessing, noise simulation, experimental results, and analysis, is described in detail.

9.1 Introduction

Deep learning has greatly aided computer vision in areas such as semantic segmentation, object detection, and object tracking. However, there are more requirements for vision models in some application areas, such as autonomous driving [7]. Though current models are capable of performing well on most tasks, they have a limitation with dirty data and fail to meet the practical standards of industrial application [2, 3, 13]. When a self-driving vehicle (SDV) is operating on the road, for instance, the complex traffic scenarios, unpredictable weather conditions, and potential sensor failure can cause pattern shifts and inaccurate object recognition. As a result, in model development, robustness and generalization are gradually brought into focus.

Current methods can be divided into three types: camera-based, LiDAR-based, and fusion methods. Camera images contain abundant information and have been shown to be effective in segmentation tasks. However, due to the limited camera photosensitivity, they would not perform well under weak or changing illumination. Previous research has explored possible solutions by utilizing specific features, such as the slender shape of lane lines and relationships between the continuous images. LiDAR is a viable alternative because it provides 3D point clouds around

X. Zhang et al., *Multi-sensor Fusion for Autonomous Driving*,
https://doi.org/10.1007/978-981-99-3280-1_9

the vehicle with information reflecting the materials and shapes of the obstacles. Unlike cameras, LiDARs can avoid the influence of lights, whereas the sparsity of point clouds limits the perception resolution.

As a complement, fusion methods can leverage advantages from multimodal data. They can fuse the source data, feature maps, or model output at different stages of the models to obtain comprehensive patterns of the target. Despite the fact that the latest research has produced ground-breaking results, few of them provide reliable explanation on the fusion mechanism or common rules in fusion model design.

We proposed that multimodal fusion be reviewed with information theory in order to create an interpretable fusion model. As most of the deep networks can be regarded as cascaded feature extraction layers, we treat each single node, layer, and pipeline and even the whole network as a communication channel [20, 26]. Therefore, the channel structure and information source determine the upper and lower bounds (channel capacity and rate-distortion bound) of the network learning ability based on the inference of Shannon's theorems. Then we suggest the balance about capacity and its distribution accordingly.

9.2 Uncertainty in Data Fusion

Dirty data has always been non-negligible trouble in perception like object detection, particularly in the field of autonomous driving and other robots. On the one hand, noise in data can cause inaccurate recognition due to the shifts in the observed pattern, necessitating a robust generalization of the models. On the other hand, the cutting-edge research leverages multimodal fusion to overcome limitations in single-sensor measurement, while few focus on the flawed data in practical applications. To tackle this issue, we propose a universal uncertainty-aware multimodal fusion framework. It adopts a multi-pipeline loosely coupled architecture to combine the features and results from multi-source data (point clouds and images in this chapter). In order to quantify the correlation in multimodal information, we further model the uncertainty in different modalities and embed it in the bounding box generation. In this way, our model can reduce the randomness in fusion and generates reliable output. Moreover, we conducted a thorough investigation on the KITTI 2D object detection dataset and its derived dirty data. Our YOLO-based fusion model has been shown to withstand severe noise interference such as Gaussian, motion blur, and frost, with only minor degradation. The experiment results demonstrate the benefits of our adaptive fusion. Our analysis on the robustness of multimodal fusion will provide additional insights for future research.

Many methods for solving the problem have been proposed, ranging from the hardware side like using more elaborate sensors to the software side such as adaptive algorithms. Some algorithms estimate the correct features for different data, including data augmentation [19], transfer learning [30], feature enhancement [3], noise estimation [29], and so on. They avoid the effects from noise in data by

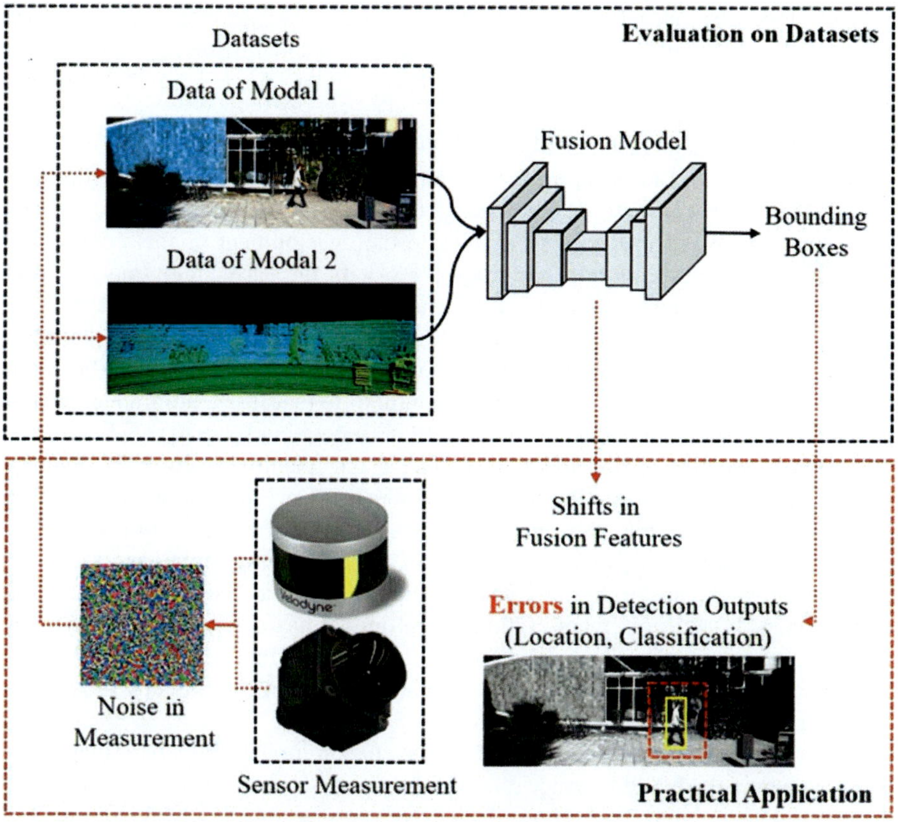

Fig. 9.1 Illustration of the measurement noise in perception models. Tests on datasets take fixed data points as input and generate statistical conclusions. While, in practical application, various environment and hardware conditions produce the potential noise in measurement, it will cause a shift in multimodal features as well as errors in the final output

grabbing the invariant features or modifying the training data distribution. However, most of these methods are developed on single-modal models that rely on the specific sensor measurement essentially. As shown in Fig. 9.1, we generally train and evaluate our models on the given datasets. Due to the limited amount of data, statistical results on the datasets cannot reflect true performance in reality, especially when sensors encounter error beyond the cases in datasets. Therefore, redundancy in the data is also a vital aspect in practical deployment.

Multimodal fusion methods have gained attention in recent years for the reasons stated above [7]. SDVs are equipped with different sensors for more complete and precise perception. On the one hand, multimodal sets can provide complementary measurements, such as cameras recording the colors and textures, LiDARs providing the 3D structure of objects, and RaDARs observing the velocity of moving targets. On the other hand, multimodal fusion can provide redundant information for stable recognition. Different sensors have specific working conditions, which means they all will fail in some situations. Therefore, we mainly discuss the adaptive strategy of multimodal fusion for SDV perception in this chapter, especially on the object detection task.

There has been a lot research into the multimodal object detection of autonomous driving since the KITTI benchmark was released in 2012 [8]. It is also recognized as a potential approach to realize robust detection. After that, various datasets and approaches were proposed to accelerate the development of this community and achieve higher recognition accuracy [4, 24]. In the early stage, model-based methods used bagging methods for result fusion [22, 28]. Individual pipelines are typically used to process different data and merge the bounding boxes to generate the final results. Latest data-driven methods mainly applied feature fusion (boosting or stacking) for a more profound information mixture that fuses multimodal data in the feature extraction or region of interest (ROI) generation stage [3, 17].

However, existing fusion methods focus on the quantified scores in standard vision tasks, with few contributing to the robustness or generalization of fusion [6]. Because of the high-dimension feature space of fusion multimodal data, fusion models are more prone to the curse of dimensionality problems. It is more challenging to extend the training data via multimodal data augmentation for the same reason. In addition, when a multimodal model has not been trained well as in Fig. 9.1, it generates greater variance with dirty multimodal data. As a result, multimodal fusion does not essentially guarantee the incrementation in performance and robustness.

Different from those learning-based feature fusion models, we claim that adaptive fusion models should leverage related information from multi-source data while avoiding noise in them. Based on this idea, we propose an adaptive fusion method with result fusion architecture. Considering that different modal data require specific operators and parameter optimization in feature processing, we adopt a loosely coupled network architecture. Multimodal data are fed to individual pipelines that are connected in the boxes filtering stage. Then we apply decision-level fusion by fusing the box proposals in the revised non-maximum suppression (NMS).

In order to achieve reliable fusion, we introduce uncertainty quantification into our model. Proper uncertainty quantification can indicate the prediction deviation of the model, so it has been viewed as a potential approach toward the interpretable neural network and an emerging method in autonomous driving [7, 14]. In our model, we predict the uncertainty for each data point. Through joint training on multimodal data, our model learns a universal uncertainty measurement that can be used as the boxes filter index in NMS. To demonstrate the benefits of our design and explore the noise effect in fusion models as well, we have evaluated the models on the KITTI 2D object detection dataset. Point clouds and RGB images were progressively perturbed to simulate multi-level dirty data. Then, we ran experiments with both raw data and dirty data. For clean data, our fusion model achieved sub-optimal but competitive results, only 0.36 mAP lower than depth model, while 4.24 mAP higher than RGB model. As for dirty data, we achieved 51.61 mAP higher than RGB model and 34.20 mAP higher than depth model on average. Our main contributions can be concluded as:

- We explored the influence of multi-level noise on LiDAR point clouds and camera RGB images and revealed the attenuation law for object detection task.
- We proposed a universal fusion model based on uncertainty, which can be implemented with different modal data and fuse their predictions adaptively.
- We conducted sufficient experiments on the KITTI dataset, demonstrating that our model has strong robustness and generalizes to noisy data beyond the train set.

Noisy data can mislead detection models because their object features are out of the distribution of the model fitting domain. Therefore, it is feasible to extend the training set to provide more diverse data or create feature filters to regularize noisy features. Data augmentation is one of the most common approaches for the former. Michaelis et al. [21] provided a benchmark to simulate multiple noise in natural environment, which is presented in Fig. 9.2. Besides, domain adaptation is another practical method. Transferring learned parameters (knowledge) among datasets can aggregate feature distributions across different data domains [30]. Khodabandeh et al. proposed to using noisy label in training to enhance the generalization [15]. Rather, researchers are concerned about whether we can directly extract and improve related features. Bijelic et al. proposed a multi-sensor model and mix features in multiple levels for mutual activation [3]. Others try to estimate the noise from the opposite perspective. Yang et al. developed a model to estimate the accuracy of laser measurement under foggy condition [29]. But most of them are

Fig. 9.2 Visualization of the simulated noise [21]. From top left to bottom right are eight common noises in nature

developed on single-modal models, which indicates that they will fail if the sensor encounters severe fault.

Most of the models discussed above avoid specific types of noisy data. However, most of these techniques do not take into account either multimodal situation or robustness of the model under various noise attacks. For example, Gaussian YOLOv3 is developed for single-modal data, UNO [25] is evaluated on semantic segmentation tasks, and Feng [6] ignores many other types of noise data from the environment. Hence, we intend to synthesize the advantages of those techniques in an uncertainty-aware multimodal object detection model and evaluate it with common types of dirty data.

9.2.1 Methodology

First, we formulate the adaptive fusion problem and describe our baseline fusion model. To improve the robustness and generate uncertainty-aware fusion strategy, we estimate aleatoric uncertainty of each bounding box of each modal via loss attenuation. Then it will be applied for boxes filtering in the NMS stage. To balance speed and accuracy, we chose one-stage object detection model YOLOv3 as our baseline, while our method should be easily extended to other classical object detection models.

The goal of noisy data object detection is to locate and classify targets from data that is disturbed with natural noise. Generally, an object detection model $\mathscr{D}(\cdot)$ takes an image, point clouds, or a group of multimodal data $X = \{X_1, X_2, ...\}$ as input, where subscripts indicate the modality. Then it returns the expected coordinates $\{x, y, w, h\}$ and categories $\{c\}$ of the targets. For noisy data detection, we assume that clean data will be measured by noisy function $\mathscr{F}(\cdot)$. Our goal is to minimize the recognition deviation while maximizing the influence of noise:

$$\min_{\mathscr{D}} \max_{\mathscr{F}} \mathscr{L}(\mathscr{D}(\mathscr{F}(X)), \{x, y, w, h, c\}) \tag{9.1}$$

As we approach the minimum, we can guarantee the generalization and robustness of our detection models over most severe noise. Specifically, for multimodal fusion models, to simplify the problem, we assume that X contains at least two modal data, namely, two types of sensors at least. The focus of this chapter is on LiDAR point clouds and camera RGB images. In addition, noise will only be added to one modal data in a single experiment, to avoid the worst-case scenario where all sensors fail and nothing works.

The YOLOv3 [23] is a modified version of the YOLO series models, achieving practicality in terms of accuracy and speed. The structure of YOLOv3 has been shown in Fig. 9.3, consisting mainly of the DarkNet-53 backbone network and the three-branch decoder. The backbone network is composed of residual blocks and convolution blocks, while each decoder branch mainly contains the convolution blocks. A convolution block comprises a 2D convolution layer, a batch normal-

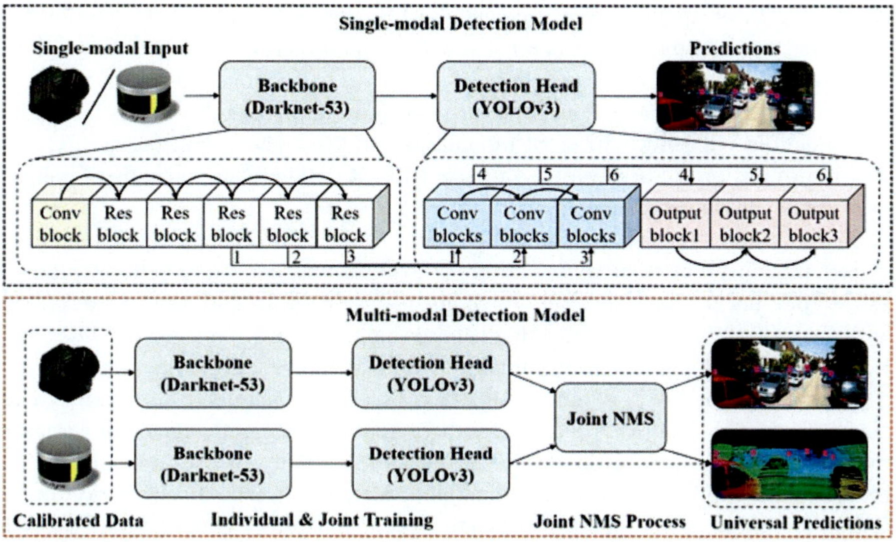

Fig. 9.3 Architecture of the proposed 2D object detection model. The top sub-figure presents the overview of YOLOv3 network [23]. For details of the network layers, refer to the original paper. The bottom sub-figure shows our loosely coupled multimodal detection model with a joint NMS for universal prediction in the 2D space. Besides, our model can be easily implemented with other backbones like SSD [18]

ization layer, and a leaky ReLU activation. A residual block comprises various residual units, and each of them has two convolution blocks and a skip-layer concatenation. Decoders process the feature maps transmitted by the former blocks and the previous branch, which will be up-sampled to realize multi-scale detection. These designs improve the detection robustness and speed of YOLOv3. To apply YOLOv3 for LiDAR point clouds, we project points onto a 2D plane as the camera image and process it as a 2D depth image.

As mentioned above, in most existing multimodal fusion network, the modalities are heavily coupled and do not consider any safety-critical environment. To weaken the interfere for fear that the modality will harm each other, we leverage parallel pipelines and late fusion techniques to combine the final outputs. The overview architecture is presented in Fig. 9.3. RGB images and the projected depth images will be calibrated in space and size. Then they will be fed to individual pipelines. Though we apply YOLOv3 for both modalities, they also accepted different combinations of models, be it 3D or 2D, one-stage or two-stage. Besides, one possible scenario is to establish communication between the two models by fusing two modal features in the middle of the models [6, 32], which may get better characteristics. However, to avoid the model non-convergence issue and two modal data uncertainty interfere with each other, our exploration of the uncertainty-aware detection model has led us to avoid this approach. For the basic fusion model, we can simply fuse all the proposed bounding boxes and filter them in the joint

NMS process or apply NMS for each modality and conduct second selection for them afterward. In the final version, this process will be revised to the uncertainty-aware multi-source NMS. The fusion model universally outputs all boxes onto the image 2D plane. Because of our loosely coupled design, the proposed fusion model is compatible with more different modalities and tasks, even with the addition of uncertainty fusion factors. But we will take LiDAR point clouds and camera images as examples in our experiments.

In our fusion strategy, we mainly consider aleatoric uncertainty, since the noises in each modality caused by sensor failure or extreme weather can be better explained by aleatoric uncertainty. Generally speaking, aleatoric uncertainty can be interpreted as noise or vague in data or label that is hard to fit.

Though aleatoric uncertainty cannot be eliminated by adding more training data, it could be reduced with additional features or views. Therefore, our strategy can also be interpreted as a type of method for reducing aleatoric uncertainty using multiple modalities. To estimate the aleatoric uncertainty of each object in each modality, we apply the loss attenuation that integrates uncertainty estimation function in the training loss function and optimize them together with neural networks. For object detection, we focus on the coordinates' regression and classification. The traditional loss function is:

$$\mathscr{L}_{NN}(\theta) = \frac{1}{N}\sum_{i=1}^{N}||y_i - f(x_i)||^2 \tag{9.2}$$

where NN represents the neural network, N represents the number of samples $\{x_i\}$, and $\{y_i\}$ represents the target output (x, y, w, h) in Eq. 9.1. In order to recognize the uncertain predictions in fusion, we expect the model can assign high uncertainty to inaccurate results and low for the rest. Then Eq. 9.2 should be redesigned as:

$$\mathscr{L}_{NN}(\theta) = \frac{1}{N}\sum_{i=1}^{N}\frac{1}{2\sigma(x_i)^2}||y_i - f(x_i)||^2 + \frac{1}{2}\log\sigma(x_i)^2 \tag{9.3}$$

where $\sigma(\cdot)$ is the variance estimation function. With loss attenuation, the model is expected to predict proper uncertainty for boxes.

Following the setting of Gaussian YOLOv3 and related information, we extend the prediction to $(\mu_x, \Sigma_x, \mu_y, \Sigma_y, \mu_w, \Sigma_w, \mu_h, \Sigma_h)$, where μ, Σ indicate the mean and variance of the target elements. Then the mean should converge to the ground truth, while smaller variances are predicted for accurate boxes, and larger variances are predicted for inaccurate boxes. Then the uncertainty can be indicated by variance. Due to the penalty during training, the samples with high uncertainty tend to generate high variance during optimization. Additionally, to fit the prediction mode of YOLOv3, we need to conduct sigmoid function for each value and scale them to the range of (0,1).

In our experiment, we examine the validation of uncertainty estimation for point cloud model and RGB model on the KITTI 2D object detection dataset. The

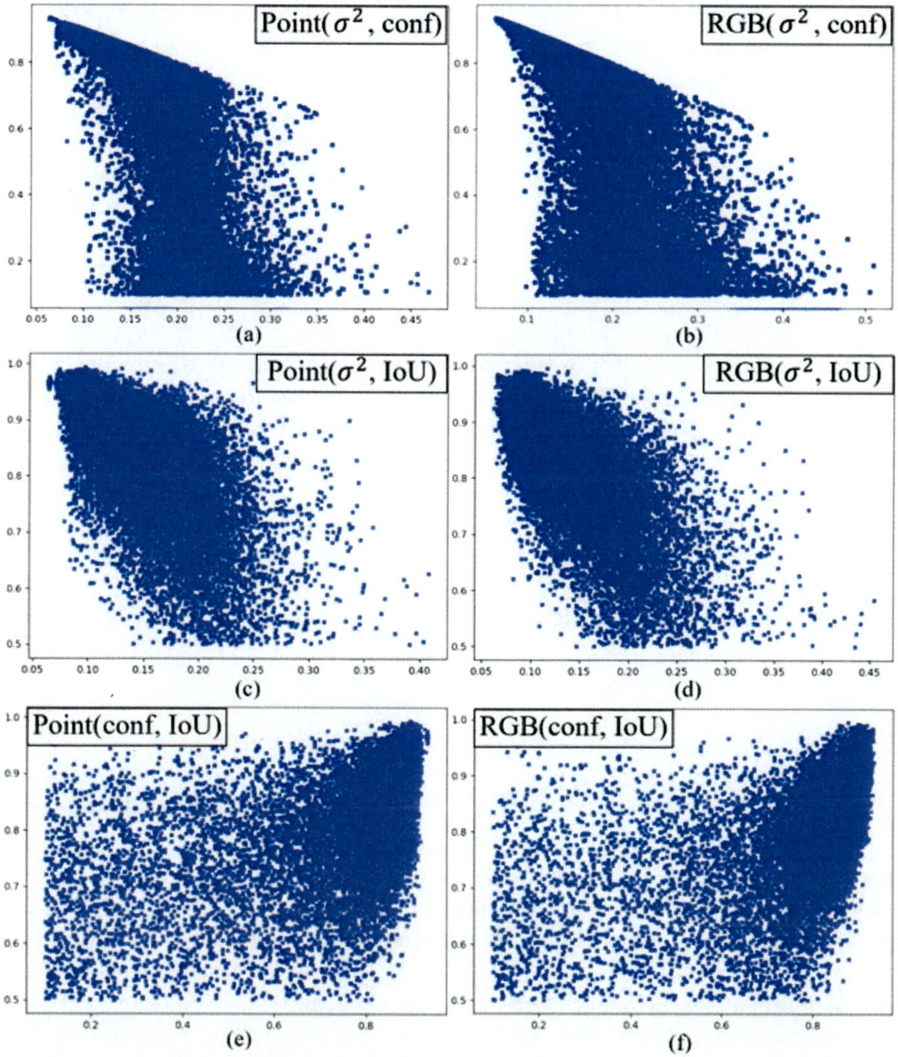

Fig. 9.4 The validity of uncertainty estimation. We visualize the relationship among uncertainty (variance σ^2, classification confidence, and localization accuracy IoU above. All of these are computed for point-based and image-based models separately as shown in the six sub-figures)

relationship among confidence (conf), uncertainty (σ^2), and localization (IoU) is visualized in Fig. 9.4. In the figures, the more linearly correlated the distributions of two variables are, the more correlated, or interpretable, they are. Figure 9.4a and b presents the joint distribution of σ^2 and conf. Their distribution is not so closer to linear, indicating that their uncertainty is less related to confidence. Comparing with this group, it is more interpretable for (IoU, σ^2) and (IoU, conf). The data points in RGB sub-figures are more concentrated than point cloud sub-figures. That means our uncertainty estimation fits RGB models better than point cloud models. This situation also applies to combination (IoU, σ^2) and (IoU, conf), indicating that the correlation between confidence and positioning accuracy is not significant. To sum up, it is reasonable to fuse and optimize the candidate boxes using uncertainty.

In this way, we model every bounding box in each object with an uncertainty value indicating the quality of the object:

$$Bbox_i \sim N(\mu_i, \Sigma_i^2) \tag{9.4}$$

Equation 9.4 is applied for the four elements $\{x, y, w, h\}$. As for our multimodal model, we have individual pipelines for each input data and correspondingly individual predicted boxes for them, which means we need to model the uncertainty in each pipeline. Because the uncertainty (variance) only reflects the properties of the boxes, we can think of these estimates as belonging to the same scale with similar meanings.

For optimization, Eq. 9.3 is revised as:

$$\mathscr{L}_x = -\sum_{i=1}^{W}\sum_{j=1}^{H}\sum_{k=1}^{K}\lambda_{ijk}\log(P(x_{ijk}^{GT}|\mu_x(x_{ijk}), \Sigma_x(x_{ijk})) + \epsilon) \tag{9.5}$$

$$\lambda_{ijk} = \frac{(2 - w^{GT} \times h^{GT}) \times \delta_{ijk}^{obj}}{2} \tag{9.6}$$

where K is the number of anchors and λ_{ijk} is set as a penalty coefficient of weight. Equation 9.5 is also applied for the four elements $\{x, y, w, h\}$ according to the specific loss function. In the setting of our result-level fusion, we predict proposed bounding boxes from two pipelines and fuse them in single NMS module, though we optimize Eq. 9.5 individually for two pipelines' output. However, uncertainty estimation is not sufficient for adaptive fusion; we further design an uncertainty-aware multi-source NMS algorithm to achieve the goal.

Non-maximum suppression (NMS) can filter bounding box according to IoU and classification score. But traditional NMS only consider classification score and ignore the accuracy of localization. Typically, when considering model uncertainty, the classification score generated by softmax layer is probable to overestimate the score. Besides, lower scored boxes may have higher localization confidence. In order to integrate uncertainty and combine the entire results of each modality in our model, softer-NMS [11] has been proposed to substitute NMS. It calculates the weighted average of all based on box-level aleatoric uncertainty and update the localization parameters for prediction. For example, x_1 is updated by:

$$x_{1,i} = \frac{\sum_j x_{1,j}/\sigma_{x_{1,j}}^2}{\sum_j 1/\sigma_{x_{1,j}}^2} \quad s.t. IoU(b_i, b) > 0.5 \tag{9.7}$$

where $\sigma_{x,i}^2$ is the aleatoric uncertainty of bounding box. All eight parameters from two pipelines will be updated with the same approach.

In the case of multi-source fusion, we have predictions from multiple pipelines. If we mix the predictions of multiple modalities directly, we will ignore the pattern correlation across different modalities and consistency within each modality, respectively. Therefore, given two thresholds t_1 and t_2, we can classify the relationship between the predictions of the two modalities A, B into three cases:

- Case 1: when $IoU(A, B) \in [t_2, 1]$, the area is activated by two modal data with high confidence.
- Case 2: when $IoU(A, B) \in [t_1, t_2)$, the area exists confusing patterns from different modalities.
- Case 3: when $IoU(A, B) \in [0, t_1)$, different modal data detect objects in different areas that are not correlated.

According to the definition above, we propose the extended softer-NMS in the algorithm to replace the joint NMS, dubbed uncertainty-aware multi-source NMS. To adapt the confidence attenuation strategy in softer-NMS, we need to merge and rerank A, B first. Confidence smoothing will be applied for all these boxes, while coordinate adjustment is modal specific. For boxes in each single-modal data, we compute the IoU with boxes from other modal detections and generate three cases. For Case 1, we focus on the highly similar boxes from all modal data. For Case 2, we apply general softer-NMS to fuse boxes adaptively. However, in the Case 3, boxes from another modality are severely off position, which are more likely to drop the localization accuracy. We ignore them to avoid potential influence from them. The iteration is based on the confidence scores and will not affect the boxes' adjustment due to its disorderliness.

9.2.2 Experiment

We select the KITTI 2D object detection dataset [8] for problem investigation and model valuation, including 7481 pairs of camera images and LiDAR point clouds. It provides over 80K 2D box annotations with several different classes: car, pedestrian, tram, van, truck, and cyclist. To avoid the influence of imbalance data distribution (we focus on multimodal robustness), we combine them into three classes: car, pedestrian, and cyclist. In our experiment, the dataset is randomly split into train, validation, and test set by the ratio of 6:2:2.

Before training, we project the 3D point clouds onto the image plane to be 2D depth images. Given a point $P_v = (x_v, y_v, z_v)^T$, we calculate:

$$P_v' = K_v[R_v|T_v]P_v \tag{9.8}$$

where K_v, R_v, T_v refer to the camera calibration matrix, rotation matrix, and translation matrix. The projected front-view point cloud depth map will be cropped to the same size as RGB images. We set the size as 128×512 in our experiment. Afterward, the value of both depth map and the RGB images are normalized to [0,1] interval.

To achieve the upper bound of noise inference in Eq. 9.1, we leverage the noise benchmark [21] and select sufficient noise levels to modify KITTI data. The benchmark contains 15 corruptions on 5 severity level, including Gaussian noise, blur, and extreme weather. Because some types of the noise cannot be applied to the point clouds like weather aspects, we chose three representative interference methods, Gaussian noise, motion blur, and frost noise, with five-level noise intensity.

With the proposed multi-source NMS, we replace the joint NMS in the fusion model Fig. 9.3 with it. Our fusion strategy can be easily changed to be based on other types of NMS or even adapted to NMS-free models. For fusion models, we first train the pipelines separately to provide stable convergence for fusion. Then we concatenate two pipelines with joint NMS or our proposed method for joint optimization. We set the IoU threshold in NMS as 0.45 for the two single-modal models, as well as set the two thresholds as 0.45 and 0.7 in our proposed models.

We have also implemented our method with SSD [18] in the experiment, which can achieve even better performance for reference. For comparison, we use average boxes' selection (AvgFusion) as the default joint NMS method. Half of the boxes from two pipelines will be randomly dropped and processed in one NMS step. In all experiments, we set the batch size as 16 and the initial learning rate as 0.0001. The experiment is conducted on NVIDIA GTX 1080 Ti with CUDA 8.0 and cuDNN v7.

Preliminary experimental results on the original KITTI dataset are shown in Fig. 9.5. In Fig. 9.5a, we compared two implementation versions based on Gaussian YOLOv3 and SSD [18], which has been revised to Gaussian SSD with uncertainty estimation. The results showed that SSD achieved better performance in all models. That means our approach can generalize to other detection models and achieve better performance by incorporating a better baseline.

However, in the subsequent experiments, we chose to use Gaussian YOLOv3 model to highlight the effect of the fusion algorithm. We present the comparison on three main categories in Fig. 9.2b. Both AvgFusion and our model achieve the sub-optimal performance between RGB and depth models due to the lack of prior information in boxes selection/fusion. However, our model has higher accuracy and approaches the optimal modal performance on all objects, demonstrating the benefits of our method on clean data and the accuracy of our uncertainty estimation that reflects the quality of predicted boxes.

We have further investigated the models' performance with noisy KITTI data for the comprehensive understanding of the robustness in multimodal fusion. First, we evaluate the performance of single-modal models with noisy data to validate the simulated noise in Table 9.1. Then, we run a similar test for our proposed fusion model in Table 9.2. NR-D means fusion with noisy RGB data, while R-ND means fusion with noisy depth data. We finally provide supplement conclusions based on the results. In all experiments, models are trained with clean data and tested with noisy data.

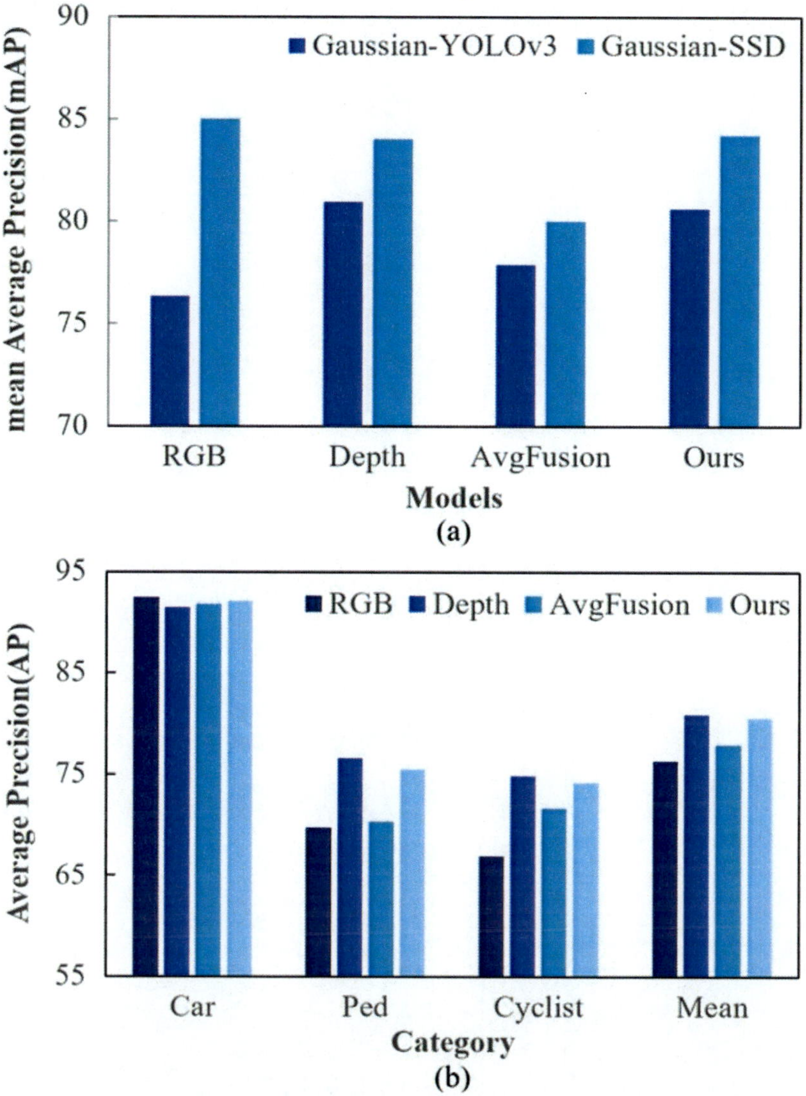

Fig. 9.5 Performance on normal KITTI dataset. (**a**) presents the comparison based on Gaussian YOLOv3 and SSD [18]. (**b**) presents the details on three categories of Gaussian YOLOv3

9.2.3 Detection Model Degradation Under Noise

The experiments show that the single-modal model is severely affected by the noisy data and the effect increases with the degree of noise disturbance. The experimental results are shown in Table 9.1. The numbers under noise name indicate the inference level. We list the mAP of the proposed model under different noise settings. Among

Table 9.1 AP for RGB/depth single-modal models under level 1–5 noisy data

Modality	Noise type		Gaussian noise					Motion blur					Frost noise				
	Category	Clean	L1	L2	L3	L4	L5	L1	L2	L3	L4	L5	L1	L2	L3	L4	L5
RGB	Car	92.39	82.13	66.31	36.81	10.04	0.39	87.68	79.01	57.36	29.83	16.32	66.19	39.30	25.34	23.17	16.38
	Ped.	69.71	41.83	21.71	6.68	0.76	0.00	40.04	19.79	7.13	0.97	0.21	21.75	4.86	1.18	1.55	0.59
	Cyclist	66.92	46.70	29.85	13.45	3.49	0.47	54.24	40.84	20.43	6.64	2.37	32.83	15.88	9.80	8.46	4.72
	mAP	76.34	56.89	39.29	18.98	4.76	0.28	60.66	46.55	28.31	12.48	6.30	40.26	20.01	12.11	11.06	7.23
Depth	Car	91.44	83.68	70.02	43.84	18.98	2.38	87.23	84.40	76.80	63.74	53.25	59.11	36.51	25.54	26.09	20.84
	Ped.	76.53	60.83	43.27	21.56	5.88	0.66	68.76	61.37	41.87	24.60	13.63	42.87	28.67	21.00	23.06	19.51
	Cyclist	74.85	64.71	53.84	36.32	18.43	2.04	67.03	60.81	48.68	36.76	27.99	53.24	44.58	37.50	37.94	34.24
	mAP	80.94	69.74	55.71	33.90	14.43	1.69	74.34	68.86	55.78	41.70	31.62	51.74	36.59	28.01	29.03	24.86

Table 9.2 AP for our fusion model under level 1–5 noisy data. NR-D/R-ND indicates one modal data is with noise

	Noise type		Gaussian noise					Motion blur					Frost noise				
Modality	Category	Clean	L1	L2	L3	L4	L5	L1	L2	L3	L4	L5	L1	L2	L3	L4	L5
NR-D	Car	92.10	90.36	88.97	87.75	87.78	88.16	90.91	89.11	87.14	86.54	87.38	88.62	88.06	87.85	87.76	87.72
	Ped.	75.48	70.52	65.64	69.40	72.03	72.77	69.16	69.68	72.30	72.48	72.68	67.24	68.43	71.38	71.71	72.31
	Cyclist	74.17	69.34	68.81	68.59	69.77	70.62	68.90	67.38	67.21	69.07	69.53	68.39	68.95	69.31	69.95	70.50
	mAP	80.58	76.74	74.47	75.25	76.53	77.18	76.32	75.39	75.55	76.03	76.53	74.75	75.14	76.18	76.47	76.84
R-ND	Car	92.10	91.40	90.60	90.62	90.73	90.87	91.97	91.44	89.79	88.17	87.40	90.31	90.41	90.62	90.72	90.70
	Ped.	75.48	74.03	73.11	68.80	65.07	63.81	73.25	69.91	65.29	62.59	60.07	72.42	68.23	66.26	66.85	67.04
	Cyclist	74.17	72.70	70.91	68.40	67.19	66.95	71.79	69.73	66.81	63.19	62.38	69.10	68.19	67.76	68.16	67.04
	mAP	80.58	79.38	78.21	75.94	74.33	73.88	79.00	77.03	73.96	71.32	69.95	77.28	75.61	74.88	75.24	74.92

them, the prediction accuracy of the RGB model shows a nonlinear decrease for the three types of noise, with mAP decreasing from 76.34 to 0.28 (Gaussian), 6.30 (motion), and 7.23 (Frost), respectively. Compared with the detection results of clean data, the accuracy of all 15 experiments decreases. Similarly, the prediction accuracy of the depth model also decreases nonlinearly with increasing noise intensity, with mAP decreasing from 76.34 to 1.69 (Gaussian), 31.62 (motion), and 24.86 (frost), respectively. The accuracy decreases in 14/15 experiments.

In addition, the sensitivity of both modal data to these noises is consistent. The severity of the noise impact, from greatest to least, is:

$$Gaussian\ Noise > Frost\ Noise > Motion\ Blur \tag{9.9}$$

In most cases, for the same class and interference level of noise, the depth model has higher detection accuracy. In all 15 sets of experiments, the depth model outperformed the RGB model in terms of accuracy, which is the same as the comparison results on clean data. The average accuracy increments are 7.74+ for car, 20.57+ for pedestrian, 22.26+ for cyclist, and 16.86+ for mAP.

Figure 9.2b shows that on clean data, our model obtains sub-optimal results close to the highest single-modal accuracy. Further, we tested the performance of the fusion model in the case subjected to single-modal data noise, as shown in Table 9.2.

The results show that the degradation of detection accuracy is smaller when our proposed RGB-depth fusion model is affected by the noise of only one modality. When the fusion model receives noisy RGB and depth data (NR-D), only 5/15 experiments show accuracy degradation. Its average accuracy changes for the three RGB noisy data are −3.40 (Gaussian), −4.05 (motion), and −3.74 (frost). mAP decreases from 80.58 to 77.18, 76.53, and 76.84, respectively. It can be seen that the noise for the RGB data has very little effect on our proposed fusion model. The effect of noise on our proposed fusion model is very small. In addition, the detection error of car is positively correlated with the noise intensity, but the other categories are not. This leads to a slight increase of mAP from the lowest point during the increasing noise intensity (in 10/15 experiments). This issue needs to be further explored in the subsequent work.

Such a situation also exists for the noisy depth data. When the fusion model receives RGB and noisy depth data (R-ND), the average accuracy decreases are −6.70 (Gaussian), −10.63 (motion), and −5.66 (frost), respectively, although there is a decrease in accuracy in 14/15 experiments. mAP decreases from 80.58 to 73.88, 69.95, and 74.92. However, there is no increase in mAP with increasing noise intensity.

Overall, our proposed fusion model is robust to single-modal data noise, and there is no substantial change in detection accuracy. Next we will analyze the gain of the fusion model on the single-modal models.

We further compare the fusion model with the unimodal model. Although the fusion model outperforms the unimodal model on noisy data, this is not incremental, as it may be degrading for the unimodal model with clean data. For example, the fusion model that received noisy data outperformed the RGB model in all tests (average mAP increase of 51.61), but not as well as the depth model and the fusion model with clean data: NRGB<NR-D<Depth<Fusion.

$$NRGB < NR - D < Depth < Fusion \tag{9.10}$$

The fusion model that received noisy data was better than the depth model in all tests (average mAP increase of 34.20), but not as good as the RGB model and the fusion model with clean data. When there is less noise,

$$NDepth < RGB < R - ND < Fusion \tag{9.11}$$

and when there is more noise,

$$NDepth < R - ND < RGB < Fusion \tag{9.12}$$

The results suggest that models with noisy data improve when they fuse clean data and, conversely, models with clean data may deteriorate when they fuse noisy data. However, there is also a smaller probability of an increase in accuracy in specific experiments, although that is difficult to explain now.

In addition, the sensitivity of fusion to RGB noise is:

$$Motion\ Blur > Frost > Gaussian \tag{9.13}$$

and to depth noise is:

$$Motion\ Blur > Gaussian > Frost \tag{9.14}$$

But overall they both have a relatively small effect. The average accuracy decreases are −3.73 (NR-D) and −7.66 (R-ND). For specific detection targets, RGB noise in multimodal data has a greater effect on car, and NR-D has lower detection accuracy than R-ND in all 15 experiments, with an average of 2.11 lower. In contrast, depth noise has a greater effect on pedestrian and cyclist, and R-ND outperforms the NR-D model in most experiments. RGB noise is more influential when noise level <3, and depth noise has greater influence when noise intensity increases.

These findings illustrate the complexity of the performance of the fusion model in the face of noisy data. For different targets, recognition tasks, data modality combinations, and noise types, fusion models may exhibit different performances. They do not show a simple linear or nonlinear change for increasing noise intensity and may even show an increase in accuracy, which requires to be further investigated.

The ablation study has been included in the comparative analysis of the single-modal and multimodal models and is therefore not listed separately. In the following, we rethink our work in terms of data noise, uncertainty, and fusion models.

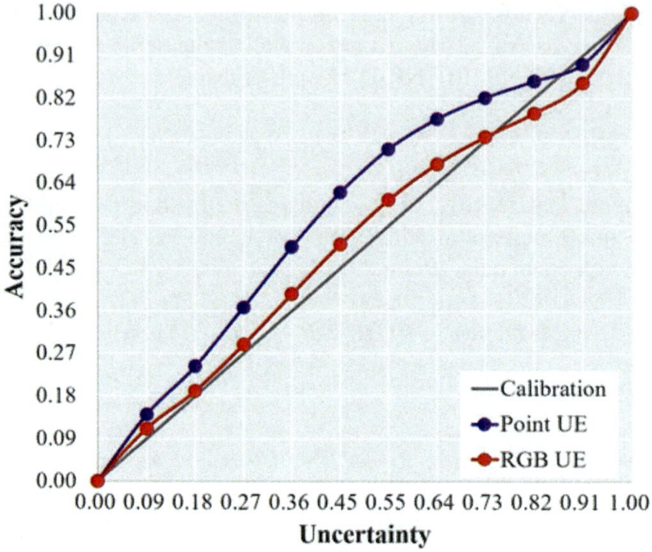

Fig. 9.6 Uncertainty calibration for the proposed model

We selected three general and suitable noises from the image noise benchmark to add to the RGB and depth data because it is difficult to capture or simulate the corresponding noise data in real scenes. However, due to the gap between data modalities, such an approach still has a large problem to fit the real error and will lead to potential bias in the experimental data. However, for the robustness of the fusion model discussed in this chapter, we mainly focus on adding noise interference of sufficient strength, so that such an error may be negligible.

Uncertainty estimation is a prerequisite of the method proposed in this chapter, and thus, the accuracy of the estimation determines the interpretability of the fusion model. Therefore, we refer to the ECE method for uncertainty calibration [16], as shown in Fig. 9.6. We model the uncertainty as a Gaussian distribution of the bounding box estimation parameters, so the magnitude of the variance corresponds to the potential range of values taken. The error between the true and estimated values can be portrayed by the distance between the curve and the $y = x$ line. The RGB uncertainty estimates are more accurate, while depth estimates have deviation, which can also lead to potential model bias.

This section have further implemented the feature fusion model with typical methods [31]. However, it is easy to have a dependence on a single modality, i.e., there is a significant primary and secondary relationship. The test results are shown in Table 9.3. We apply weighted feature addition at seven layers in the two-pipeline network. The 2×7 weight parameters will be optimized during training. When trained without any limitation, the network tends to assign more weights for the depth branch, which will seriously affect the robustness of the model. The fusion weight is then adjusted in training by using imbalance data dropout (10% for RGB

Table 9.3 Modality bias in feature fusion. L1–L7 indicate the seven fusion layers from front to back in the DarkNet in YOLOv3. NR and ND represent noisy RGB and noisy depth. w/ and w/o indicate whether we limit the fusion weights in training

Exp.	Modal	L1	L2	L3	L4	L5	L6	L7	NR	ND
w/o	RGB	0.3	0.6	0.7	0.7	0.5	0.5	0.1	83.2	19.8
	Depth	0.7	0.4	0.7	0.7	0.7	0.9	0.7		
w/	RGB	0.6	0.5	0.8	0.8	0.7	0.4	0.3	77.6	39.5
	Depth	0.3	0.1	0.2	0.1	0.2	0.8	0.7		

and 15% for depth) and weight limitation (upper to 0.7 in the first five stages). The adjustment improved the performance slightly. But it is far from enough, and adaptive balance fusion remains an open problem.

9.3 Information in Data Fusion

There has been a recent growing interest in utilizing multimodal sensors for accurate lane line segmentation. In this chapter, from information theory perspective, we put forward a novel multimodal fusion architecture and demonstrate its practical utility using Light Detection and Ranging (LiDAR) camera fusion networks. The forward propagation is thus equal to the information transmission in the channels. In particular, for the very first time, we get the multimodal fusion network developed as a joint coding model, where each single node, layer, and pipeline are represented as a channel. Then, we can conduct a qualitative and quantitative analysis of the impact of various fusion approaches. We believe the ideal fusion architecture is related to the essential capacity and its allocation based on the source and channel. To verify this multimodal fusion hypothesis, we progressively list a series of multimodal models based on the suggested fusion techniques and evaluate them on the KITTI and the A2D2 datasets. Our optimal fusion network achieves 85%+ lane line accuracy and 98.7%+ overall. The performance gap among the models will provide information to help future's research on the creation of ideal fusion algorithms for the community of deep multimodal learning.

In addition, to validate our inference, we undertake comparative tests, and based on which, a novel fusion network is proposed. On the one hand, we explore the respective effect of different factors in fusion, including the fusion stage and network structure. On the other hand, we test the impact of information source through modality loss experiments. According to the results on the KITTI and A2D2 benchmarks, our ideal model gains 8.51% lane line accuracy or 7.6 F2 score over the baseline. It also performs better when a single modality is removed. As shown in Fig. 9.7, the result demonstrates the benefits of our fusion strategy. Additionally, our lightweight networks are constructed end-to-end that can achieve real-time segmentation at over 64.9 FPS with lane line accuracy of 85%+ and total accuracy of 98.7%+ on both datasets.

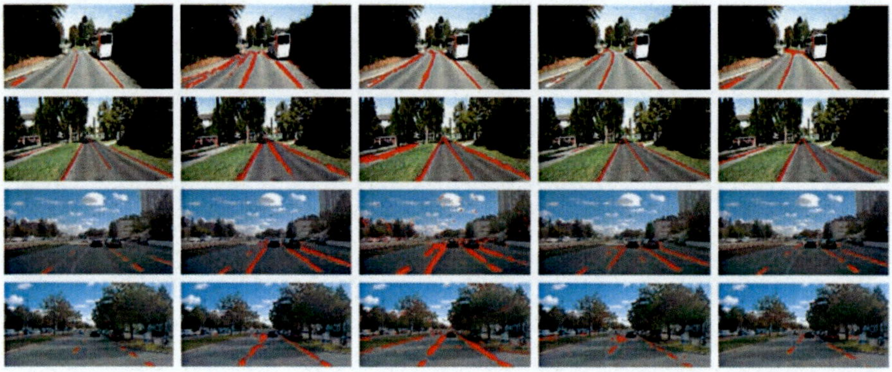

Fig. 9.7 Comparative results of two state-of-the-art models (SCNN, LaneNet) and ours: the first two rows are tested on the KITTI dataset and the rest are tested on the A2D2 dataset

In conclusion, our main contributions are:

- We propose to review the multimodal network from the vantage point of channel model and provide reasonable illustration for fusion.
- We provide qualitative and quantitative analysis on different factors that can impact the fusion networks based on information theory.
- We present a novel multimodal fusion network for lane line segmentation tasks that approaches state-of-the-art performance.

In the following sections, firstly, we will review the recent progress in lane line segmentation, multimodal fusion, and information theory, and secondly, we will illustrate how to model and analyze the network from the framework of channel model and discuss the fusion strategy referring to Shannon's theory. Moreover, we will conduct quantitative analysis on the channel and source in the experiments and ablation study.

Information theory, proposed by C. Shannon, studies the information process on the basis of the entropy-based quantification of information. By quantifying the level of signal uncertainty, the information entropy reveals the limits on signal processing and communication operations. Other significant measurements of information theory include mutual information, channel capacity, and error exponents. In fusion network research, however, they are uncommon. MacKay et al. have discussed modeling a neuron or a network as a channel. N. Tishby et al. proposed the information bottleneck theory (IB) with a variational principle to address signal processing issues [27]. In 2015, Tishby and Zaslavsky further revealed the mechanism deep learning model using the IB. Their experiments with a multi-layer perceptron showed that the network tends to capture related information first and combines them later. Though the result has not been generalized well to more complex CNN models, it inspires us to review the fusion model as a channel and the fusion process as information gain.

9.3.1 *Multimodal Fusion Within the Context of Information Theory*

To reveal the relationship between the information theory and multimodal fusion, we first model the network as a joint coding channel, and then Shannon's theorems can be used to reveal the fusion mechanism. We will then present a practical fusion strategy based on our analysis.

As shown in Fig. 9.8, a fundamental communication system consists of five parts: information source, transmitter, channel, receiver, and destination. To represent a deep network as a channel, the five components are the source to measure, the encoder (including sensors and encoders), the hidden layers (channel), the decoder, and the desired output. Generally, we can also view the learning as joint source-channel coding, in which the encoder, channel, and decoder are combined. In this way, we can further look into every single layer for a more precise analysis on the

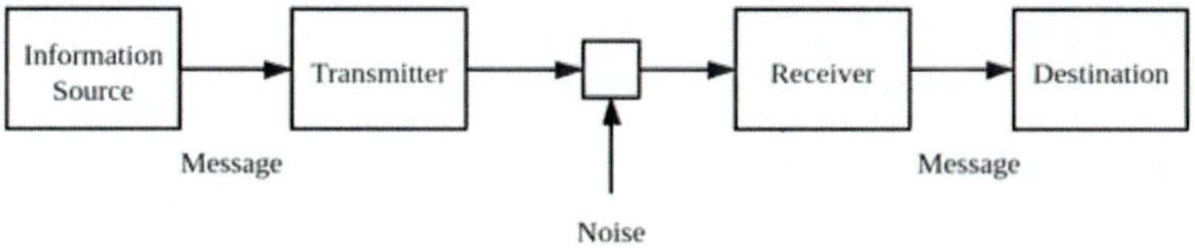

(a) The Communication System in Shannon's Paper

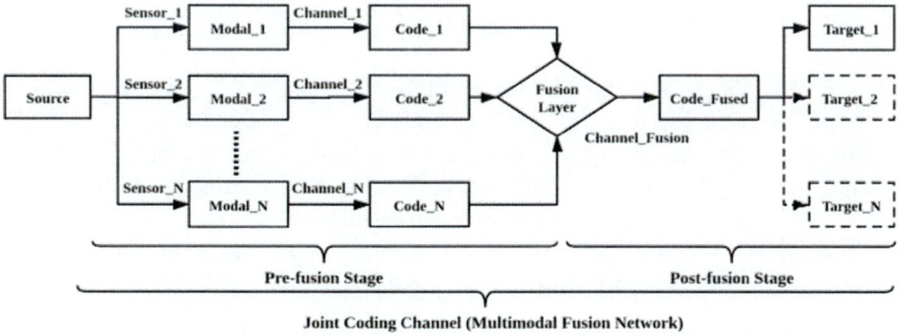

(b) Model The Fusion Network as A Joint Coding Channel

Fig. 9.8 The traditional communication system and the multimodal network: For the fusion model, different modal data are processed in separate channels, fused in the fusion channels, and decoded to be the desired output, which is the projection of the target in one modality fusion. Specifically, each layer or pipeline is a channel, and the entire network is a cascaded channel.

To make it clear, we hereby define that the number of output channels in each layer equals to the code length n, the features equal to the codes, and the learning ability equals to the channel capacity. The code set M therefore contains all the possible features, and the rate of code $R = \frac{\log_2|M|}{n}$ represents the efficiency of the code, that is, the redundancy of network. Researchers have also discussed the estimation of the information entropy and mutual information [1, 9, 27], but they have not yet devised a method that is optimal for all networks. Thus, we assume that the networks we mentioned have a potential entropy, which is not used directly in the following. Though many details on the definition might have been missed, we can conduct basic qualitative analysis and quantitative comparison for the fusion problems in multimodal lane line segmentation.

In addition, in the case of multimodal fusion network, the fused features and information are generated from the joint coding in the fusion channels. As the multimodal data and features in the coding originate from the individual $sensor\_i$ and corresponding $channel\_i(i \in \{1, 2, ..., N\})$, multimodal learning is actually a form of parallel joint coding.

As the foundations of the information theory, Shannon's theorems reveal the relationship between coding and the ability to transmit information for certain source and channel models. They also imply that the optimal multimodal fusion architecture is determined by both the data quality and the network itself. Initially, we briefly review Shannon's theorems as below:

Theorem 1 Define R as the code rate and C as the channel capacity. For any discrete memoryless channel, if $R < C$, then R is achievable. Conversely, if $R > C$, it is not achievable.

Theorem 2 The channel capacity C is:

$$C = B \times \log(1 + S/N) \tag{9.15}$$

where B is the bandwidth of channels and S/N is the signal-noise ratio (SNR).

Theorem 3 The rate-distortion function for an i.i.d. source X with distribution $p(x)$ and bounded distortion function $d(x, \hat{x})$ is equal to the associated information rate-distortion function. Thus,

$$R(D) = \min_{p(\hat{x}|x):\Sigma_{(x,\hat{x})}p(x)p(\hat{x}|x)d(x,\hat{x})\leq D} I(X; \hat{X}) \tag{9.16}$$

is the minimum achievable rate at distortion D.

In summary, these theorems focus on the source coding and the transmission in channels. Theorem 1 indicates that the rate of code is bounded by the channel capacity, which is determined by channel bandwidth and SNR. Theorem 3 provides a lower bound for the rate. Notice that the channel capacity only reflects the performance of the channel, while rate-distortion targets the source. By combining them, we found that for the given distortion D, a code is achievable if and only if the code, channel, and source satisfy the inequality:

$$R(D) \leq R \leq C \tag{9.17}$$

Single-Modal Learning

These theorems can be applied directly to deep learning, which presents the limits of learning (channel capacity) and compressing (rate-distortion). To avoid the complicated calculation of network capacity like MacKey et al. and simplify the problem, we will perform qualitative analysis on the model instead. Specifically, for single-modal learning, the rate-distortion function has been determined based on the dataset, and the channel capacity can be adjusted automatically or manually during training. Moreover, we can rewrite Eq. 9.17 informally as below:

$$D = D(R) \geq D(C) \tag{9.18}$$

Equation 9.18 shows that by properly raising the capacity and rate of code, the network can achieve less distortion. But as the source information is limited, the enhancement will also be subtle.

Multimodal Learning

The case in multimodal learning will be slightly different from the single-modal, especially the channel capacity and source distortion. Assume that the object has a high-dimensional feature space and different sensors only observe its subspace in specific modalities. As shown in Fig. 9.8, a basic fusion model can be divided into two stages: pre-fusion and post-fusion. In pre-fusion, it holds the inference for single-modal learning in each pipeline. But in post-fusion, the input is the combination of feature subspaces, and the $channel\_fused$ is the extension of sub-channels in pre-fusion. This identifies the individual rate-distortion function and required capacity for each modality, which will further influence the optimal bandwidth (weight) allocation of the sub-channels in fusion, with the balance between pre-fusion and post-fusion. In addition, the capacity is determined by the bandwidth and SNR. The bandwidth measures the size of feature subspace, whereas SNR reflects the proportion of effective data transmitted through the network. The SNR of data is the SNR of sensor measurement, which is also a channel.

In conclusion, the channel model and Shannon's theorems provide a novel perspective to review the deep multimodal learning. Furthermore, they also imply that the optimal fusion architecture depends on three balances: the capacity allocation for sub-channels, the division of pre-fusion and post-fusion, and the conflict between information distortion and channel redundancy. In essence, the fusion model will strike a balance by improving the channel capacity or using a better information source. Then it will be able to develop novel information-driven fusion models based on our deduction.

Information-Driven Fusion Strategy

Toward the optimal LiDAR-camera fusion in lane line segmentation, we propose a novel information-driven multimodal fusion strategy with a two-stage analysis on the information source and network.

First, we estimate the required channel capacity based on the information source including the capacity in the overall model and its allocation for each modality. According to Theorem 1, uncertain data (with a high entropy) transmit more information via the channel and require higher rate of code, namely, more capacity. Supplementally, the divergence between the training set and test set contributes to the uncertainty. It's a same case as the capacity allocation for the pre-fusion pipelines. Those data with higher SNR, for example, a well-captured image compared with the sparse point clouds, are supposed to occupy larger capacity in fusion.

Besides, the ideal fusion also depends on the data SNR. In reality, when the bandwidth falls during the fusion process, post-fusion forces the model to "forget" noise in joint optimization and improves its generalization to subsequent problems. Conversely, using more pre-fusion would help to adjust the training set as it adopts more information from the source. Therefore, when the uncertainty increases, like using small training set or hard test set, earlier fusion will perform better.

Then, we can adjust the channel architecture based on these estimations, and all of them will essentially change the capacity through bandwidth or SNR. For the capacity allocation, it is practical to apply adaptive bandwidth/weight in fusion, for instance, the adaptive weight estimator, fully linear connection, depth-wise convolution, or other individual processes for different modal data. But for the overall capacity, we are contemplating the use of cascaded channel structure, including tandem channels (like attention model and cascaded detectors) and parallel channels (like the Y-shaped fusion in Fig. 9.8). Both types aim to increase the capacity with longer codes: the tandem focus on the depth of network and the parallel target the width. As the optimal model can be the combination of them, we integrate the road segmentation and our model into multi-task learning and combine the early and middle fusion to extend the capacity, which comprises both tandem and parallel structure. Details are referred to the model section.

Theorem 4 A q-ray $(n, |M|, d)$ code C can correct t errors if and only if:

$$d(C) \geq 2t + 1 \tag{9.19}$$

where n is the code length and $d(C)$ is the minimal distance among codes.

The theorem shows that a model can achieve better performance of the source with longer code, but at the expense of performance robustness. It not only informs the conflict between essential capacity and redundancy but also indicates a possible approach to compare the contribution of different modalities by computing their error correction ability.

Beyond these, although the qualitative analysis method has been proposed previously, we will still be unable to provide guidance to a precise fusion model if there are no quantification tools supported. Instead, it provides guidance for the architecture. Specific fusion models will be introduced in the following sections.

9.3.2 *Multimodal Models*

To better demonstrate our research, we will conduct a series of experiments to uncover the impact of different fusion stages and methods. We first introduce baseline models and then move to multimodal fusion methods.

Single-Modal Network Baseline
We consider an encoder-decoder network as our baseline not only for its popular model architecture but also for its ease in ablation investigation, in which we can focus on the fusion approach rather than the network itself. Besides, the limited amount of multimodal lane line data requires a small model. Therefore, we select U-Net as our baseline as shown in Fig. 9.9. It comprises four blocks in the encoder and five in the decoder, in which two are ResNet-34 blocks. The last four layers use transposed convolutions and the rest are convolutional blocks. All convolutional blocks have a batch normalization layer and a ReLU layer following the convolution layer, and all kernel sizes are 3 × 3. Each block in the encoder is connected to its corresponding blocks in the decoder through a dashed line, which concatenate their respective outputs in order to correct the feature maps.

Point Cloud Completion
The sparse reflectance is obtained by projecting the front-view point clouds from the front view onto the imaging plane of the camera in order to align with the RGB images. But the ratios between point clouds and images are around 1.5% in KITTI and 0.4% in A2D2. Therefore, point cloud completion can improve the channel capacity with higher SNR. We apply the k-NN search for interpolation: search three nearest points for each blank pixel, and count the weighted average based on the normalized pixel distance as the result. To better decrease noise caused by reflectance attenuation and utilize the height information as an additional filter

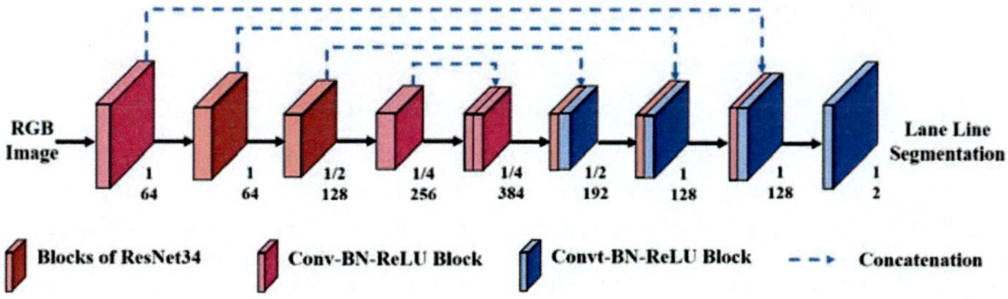

Fig. 9.9 Our single-modal baseline model (V1): It accepts an RGB image as input and outputs a 256 × 128 binary map

condition, we stack the height and distance value on the reflectance to compose pseudo-three-channel images.

Fusion at Different Stages

In order to find the ideal fusion architecture for multimodal lane line segmentation, we have progressively constructed and compared multiple fusion models, including the early, middle, and late fusion as shown in Fig. 9.10. In the point cloud pipeline,

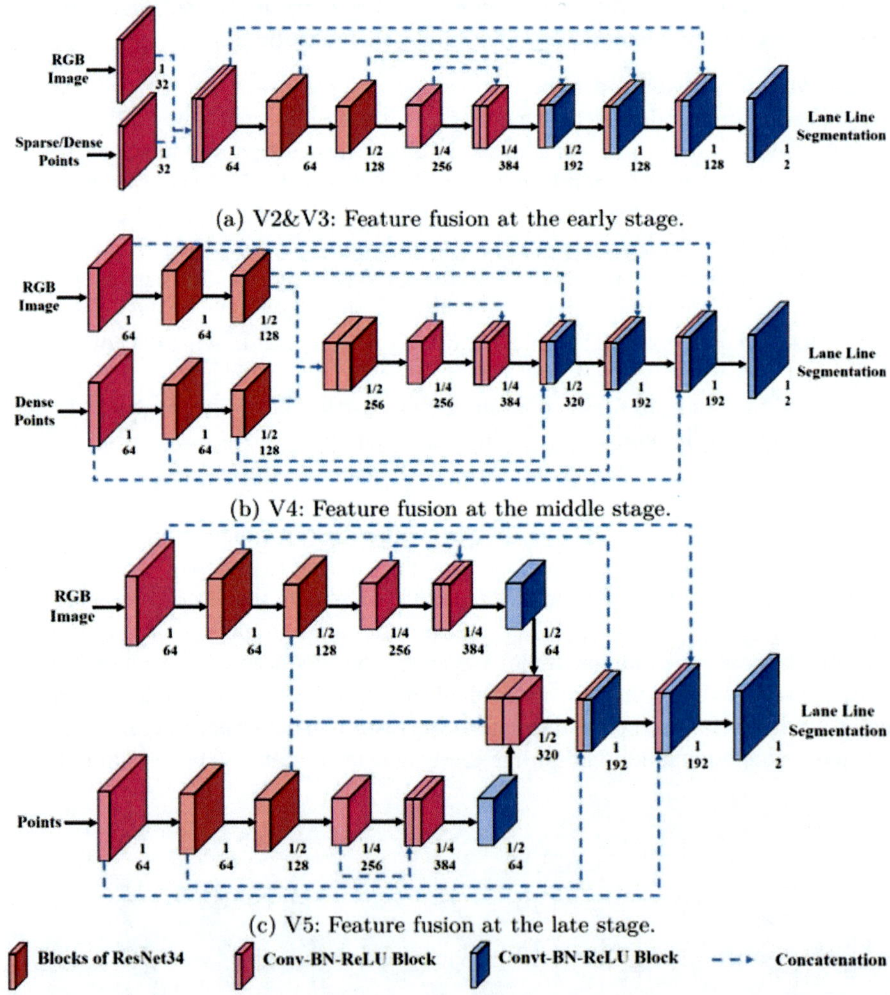

Fig. 9.10 Feature fusion at three stages: the top model (V2 and V3) fuses at the early stage; the middle one (V4) fuses at the middle stage; and the bottom model (V5) conducts fusion in the decoder. Among these models, V2 uses sparse point clouds, and V3, V4, and V5 use completed point clouds. All the three fusion belong to the "middle fusion"

we use convolutional layers rather than ResNet blocks to achieve the capacity allocation in V4 and V5. Features from two modalities will be concatenated to double channels. Though adaptive bandwidth fusion as mentioned previously would perform better, we apply convolution after concatenation for lightweight models. Besides, instead of paying attention to the potential complex fusion operators, we adopt the general approach by concatenating features from different inputs. Specifically, we use early, middle, and late fusion to indicate data fusion, feature fusion, and result fusion. Therefore, we concatenate preprocessed point clouds and images after the first convolution layer to build V3 (early fusion) and use skip connection to combine features from the point and image pipeline in V4 (middle fusion). As for result fusion, the communication of multimodal data in V5 is placed at the later stage of the model. For more details of the model parameter, refer to Fig. 9.10 and experiment section.

Multi-Task Learning

Road segmentation is considered as an efficient way to extract prior knowledge and gain more capacity for lane line. Typically, lane lines exist in the driving area; thus, filtering the background in images and point clouds can increase the SNR. Moreover, adding a relevant task equals to adding the code length, which means a larger feature subspace. Apart from the capacity, with Theorem 4, longer codes can also enhance its error correction capability. However, the interference among sub-tasks will impact the convergence rate in training. As shown in Fig. 9.11, we design a two-branch decoder block to conduct multi-task learning, which can be easily extended to other tasks. The block utilizes the last three decoder blocks in

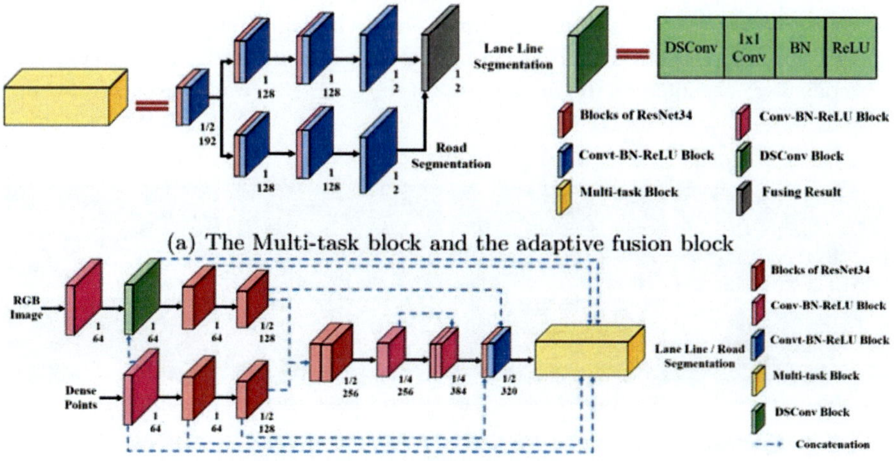

(a) The Multi-task block and the adaptive fusion block

(b) The Final Model (V6) with Early & Middle Fusion

Fig. 9.11 The multi-task block and the adaptive fusion block (DSConv): The gray layer combines the results from two branches each branch and combines the result to revise the prediction of the main task, lane segmentation. The output of the lane is:

$$P(X \in lane|road) = P(X \in lane) \times \{k + (1 - k) \times P(X \in road)\} \quad (9.20)$$

where k is a trainable parameter and the equation helps to decide to what extent the lane line prediction relies on road. According to the performance in V3–V5, we add the multi-task block for V3 and V4 to build V3r and V4r.

Adaptive and Multi-Stage Fusion
We use a fusion block for adaptive fusion across multimodal feature channels. It contains a depth-wise convolutional layer, a 1×1 convolutional layer, a batch normalization, and an activation layer. Therefore, important features will dominate training after the 1×1 convolution. We embed these methods in the V3r model to build V3r+. Besides, the final version (V6) are implemented with fusion at multiple stages based on the performance of previous models. Actually, V6 fuses at the early and middle stages simultaneously. This common X-shaped architecture can further enlarge the capacity and join the features of V3 and V4.

9.3.3 Experiment

Data Augmentation
To alleviate overfitting problems with small datasets like KITTI, we use a variety of methods to enhance the dataset. These methods primarily comprise geometric transformations—such as perspective transformations, random rotations, and flips—as well as pixel-level enhancements, such as adjustments to brightness and contrast. Furthermore, target interference techniques—like Gaussian noise, random clipping, and the random erasure of partial lane lines—are also utilized. All augmentation methods were executed on the images except the geometric translation (Fig. 9.12).

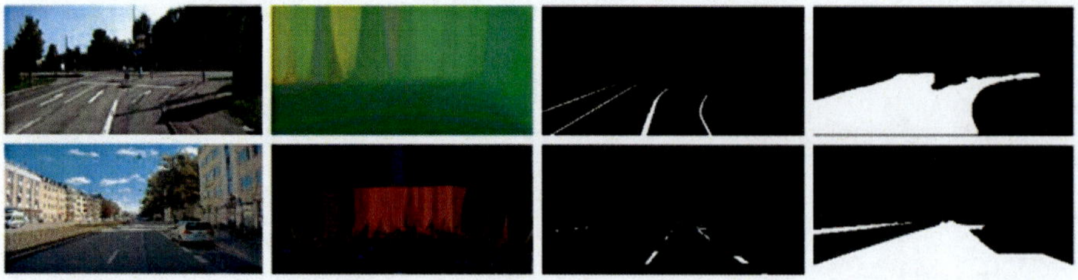

(a) RGB image (b) Point Clouds (c) Lane Label (d) Road Label

Fig. 9.12 Some examples of the KITTI (top row) and the A2D2 dataset: point clouds are the composed output of k-NN completion

Data Preprocess

To integrate LiDAR point clouds and RGB images into the same network, projection and value normalization are required in data preprocessing. To project the point clouds onto the image plane, given a point $P_v = (x_v, y_v, z_v)^T$, we calculate:

$$P_v' = K_v[R_v|T_v]P_v \tag{9.21}$$

where K_v, R_v, T_v refer to the camera calibration matrix, rotation matrix, and translation matrix. Then the projected front-view point cloud reflectance map is then reduced to the same dimensions as RGB pictures, which is 128*256. After that, both reflectance map and the RGB images will then have their values normalized to [0,1] interval.

Optimization

All models are trained with the AdamOptimizer with a cyclically decaying learning rate lr:

$$lr = 2^{\lfloor epoch/50 \rfloor} \times 0.8^{\lfloor epoch/10 \rfloor} \times lr_0, \; lr_0 = 0.0001 \tag{9.22}$$

Adaptive Weighted Loss

It is common to use weights in classification, but most methods rely on prior knowledge of the dataset or adjust training parameters. during training. Inspired by He et al. [10], we use the $torch.nn.NLLLoss2d$ in PyTorch and set the weight to be the inverse ratio of two classes in the previous prediction. However, sometimes a branch will dominate training and cause unexpected weights, like training with multi-task. For better convergence, we set the weights as (0.5,0.5) in the first 20 epochs.

Hardware and Software

Our models are trained on a single GPU, GTX 1080 Ti with 11G RAM, and E5-2678v3 CPU. Besides, we use the following software setup: Ubuntu 16.04 64-bit operating system, Python 3.6, GCC 5.4.0, and PyTorch 1.10 with CUDA 9.0 hardware acceleration.

We focus more on $lane\ linerecallandcalculateitaslaneaccuracy$ $(LAcc)$. We also consider the $F2\ score$ to balance in case the network overfits any class and count the mean recall on both class as the $mAcc$. In the road segmentation, we apply the same metrics:

$$precise = \frac{TP}{TP + FP}, \; LAcc = recall = \frac{TP}{TP + FN} \tag{9.23}$$

$$F2 = \frac{(1 + 2^2) \times precision \times recall}{(2^2) \times precision + recall} \tag{9.24}$$

$$Acc = \frac{TP + TN}{TP + TN + FP + FN} \tag{9.25}$$

$$mAcc = \left(\frac{TP}{TP + FN} + \frac{TN}{TN + FP}\right)/2 \tag{9.26}$$

In Eq. 9.23, TP, TN, FP, FN refers to true positive, true negative, false positive, and false negative.

We compare our models (V1–V6, V3r, V4r, V3+) with the selected subset of the KITTI and A2D2 datasets with SCNN and LaneNet, two of the most prominent models in the field. Except for SCNN and LaneNet, which load the pre-trained VGG-16 weights to expedite learning, all models initiate training from scratch. To be fair, we train SCNN and LaneNet for 10,000 iterations (equal to 175 epochs, since they have stopped optimization after 5000 iteration) and our models for 200 epochs due to the computing resource limitation. Consequently, fine-tuning or additional training can improve their performance. The training record of our models on the KITTI dataset is shown in Fig. 9.13. All models achieve the same performance after around 150 epochs without significant gap among different fusion stages, but adding the multi-task learning will hinder the fitting due to the message conflict in channels. We also notice that V3+ and V6 accelerate after 20 epochs which verifies the influence of adaptive fusion on capturing important features.

According to Table 9.4, the accuracy of the majority of models on both datasets exceeds 98.7%. But all F2 scores are lower than 65.0, which can be attributed to the extreme imbalance between the lane line and the background (almost 98% background). Inaccurate prediction on a small number of pixels can result in a substantial variance in recall and precise, which can be solved by using precise post-processing, let alone the deviation caused by annotation. As for LAcc, we found that the earlier it fuses, the better it performs. Besides, V3r and V4r get higher scores than V3 and V4, indicating the benefits of multi-task learning. By comparing the results of KITTI and A2D2, we found that the point cloud completion can even weaken the model, as it can be caused by the low quality of A2D2 LiDAR for lane line segmentation. The V6 model, which is built according to the results of V3V5, outperforms other versions in almost all tests.

Finally, the model speed is determined by the forward segmentation network cost. Except for the V5 model, all our other models can achieve a minimum of 62 FPS with less than 4G memory on GPU. As shown in the table, our lightweight models can approach the state-of-the-art performance on both KITTI and A2D2 datasets and meet the requirements of autonomous driving. Notice that our models are based on a fundamental framework and they are supposed to bear with more optimal architecture like knowledge distillation [12] and multi-objective particle swarm optimization [5] in the future work. To further clarify what contributes to the performance, more experiments are conducted on the KITTI and augmented KITTI datasets with our models in the following ablation study.

In this section, we further conduct quantitative analysis on the proposed strategy. First, we discuss the contribution of different factors in fusion from channels'

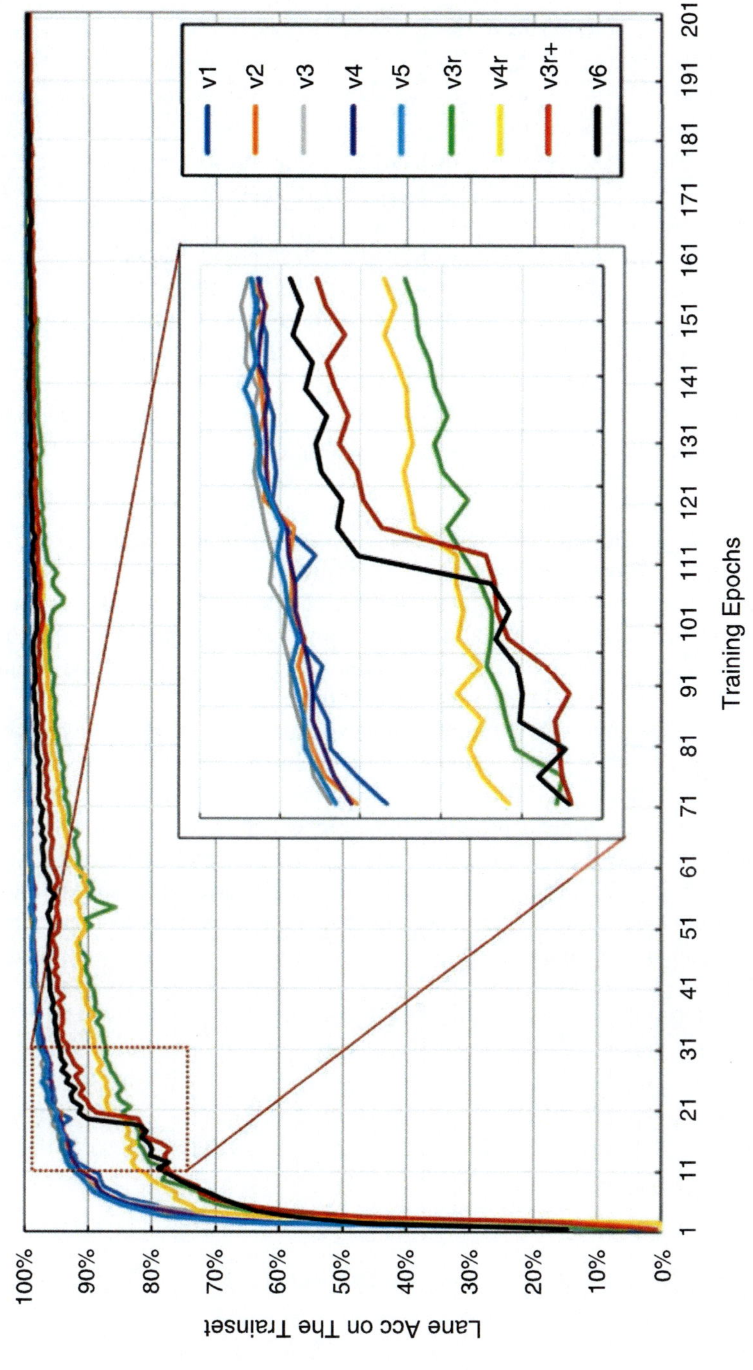

Fig. 9.13 The convergence process in training: The curves show the accuracy for lane on the training set. V1, the baseline; V2–V5, feature fusion at early, middle, and late stages; V3r and V4r, multi-task training version for V3 and V4; V3r+, refined version of V3r by adding adaptive blocks; V6, the full and final version

Table 9.4 The results on the KITTI and the A2D2 datasets: E/M/L denotes the early/middle/late fusion

			KITTI (383)				A2D2 (788)				
Model	Fuse	Size (M)	LAcc	Acc	mAcc	F2	LAcc	Acc	mAcc	F2	FPS
LaneNet	/	556	70.97	95.78	83.46	33.7	78.83	97.30	88.12	41.5	69.1
SCNN	/	556	84.68	98.29	91.55	58.4	80.93	96.06	85.54	34.2	14.4
V1	/	25	82.38	98.71	90.63	62.2	82.55	98.43	90.54	54.3	**73.3**
V2	E	25	82.50	98.70	90.68	62.0	81.80	98.68	90.29	57.4	68.1
V3	E	25	84.90	98.73	91.89	**64.0**	81.55	98.79	90.22	59.6	68.9
V4	M	31	84.46	98.72	91.67	63.5	79.85	98.80	89.38	58.6	**72.4**
V5	L	31	82.66	98.77	90.80	63.6	70.28	99.10	84.79	57.5	55.0
V3r	E	27	**86.21**	98.64	**92.49**	63.9	**86.99**	98.02	**92.54**	51.5	63.2
V4r	M	35	83.47	98.74	91.18	63.3	83.32	98.59	90.51	56.4	63.4
V3r+	E	27	**85.92**	98.67	**92.36**	63.3	84.90	98.69	**91.74**	59.3	62.3
V6	E+M	35	**86.48**	98.76	**92.69**	**64.9**	85.13	98.86	**92.04**	**61.9**	64.9

Bold values highlight the significant improvement in performance

perspective and then compare the robustness in cases where the models lose one modality in practical usage, which also verifies our inference.

Separately training and evaluating our models on the KITTI and its enriched dataset allows us to investigate the influence of source and channel. Their performances in LAcc, mAcc, and F2 score are listed in Table 9.5. Based on the result, V3r, V3r+, and V6 are the best when training with the original KIITI data, and V3r, V4r, and V6 lead in the augmented training. However, the fusion models can be even worse in some cases: the V5 is weakened on the augmented dataset, and V2–V5 perform poorly when using KITTI for training and the augmented data for testing. The instable performance reflects the potential dependence on information source. Apart from this, we assume every factor are linearly independent and quantify their effect by solving Eq. 9.27 by using least squares, obtaining the final result in Table 9.6. Note that the value does not represent the absolute gain or loss in fusion, but rather the proportional contribution.

$$\begin{bmatrix} 1\,0\,1\,0\,0\,0\,0 \\ 1\,1\,1\,0\,0\,0\,0 \\ 1\,1\,0\,1\,0\,0\,0 \\ 1\,1\,0\,0\,1\,0\,0 \\ 1\,1\,1\,0\,0\,1\,0 \\ 1\,1\,0\,1\,0\,1\,0 \\ 1\,1\,1\,1\,0\,1\,1 \end{bmatrix} \times \begin{bmatrix} LiDAR \\ dense \\ early \\ middle \\ late \\ road \\ adaptive \end{bmatrix} = \begin{bmatrix} Result\ of\ V2 \\ Result\ of\ V3 \\ Result\ of\ V4 \\ Result\ of\ V5 \\ Result\ of\ V3r \\ Result\ of\ V4r \\ Result\ of\ V6 \end{bmatrix} \tag{9.27}$$

According to Table 9.6, the fusion network has different performances based on the train/test case. The raw data is a subset of the augmented data, with the cases A-A and A-K bearing a smaller train/test distance, while K-K and K-A a larger gap.

Table 9.5 A comparison of different fusion strategies: "dense" denotes the point clouds completion, and "adap" is the adaptive block

		Fusion						KITTI			KITTI (Aug)		
Data	Model	Dense	Early	Mid	Late	Road	Adap	LAcc	mAcc	F2	LAcc	mAcc	F2
KITTI	V1							82.38	90.63	62.2	82.55	90.54	54.3
	V2		✓					+0.12	+0.05	−0.2	−1.86	−1.21	+3.7
	V3	✓	✓					+2.52	+1.26	**+1.9**	−1.00	−0.32	+5.3
	V4	✓		✓				+2.08	+1.04	+1.3	−2.70	−1.16	+4.3
	V5	✓			✓			+0.28	+0.17	+1.4	−12.27	−5.75	+3.2
	V3r	✓	✓			✓		**+3.83**	**+1.86**	+1.7	**+4.44**	**+2.00**	−2.8
	V4r	✓		✓		✓		+1.09	+0.55	+1.1	+0.77	−0.03	+2.1
	v3r+	✓	✓			✓	✓	**+3.54**	**+1.73**	+1.1	+2.35	+1.20	+5.0
	V6	✓	✓	✓		✓	✓	**+4.10**	**+2.06**	**+2.7**	+2.58	**+1.50**	**+7.6**
KITTI (Aug)	V1							85.29	92.14	66.0	79.45	85.84	56.4
	V2		✓					−2.28	−1.04	+1.3	−1.10	−0.25	−0.3
	V3	✓	✓					−0.05	−0.25	**+3.7**	+3.15	+3.09	**+6.1**
	V4	✓		✓				+0.41	+0.27	+2.1	+2.04	+0.87	+3.0
	V5	✓			✓			−3.74	−1.75	+1.3	−2.60	−0.63	+2.4
	V3r	✓	✓			✓		**+3.75**	**+1.62**	−3.6	+5.47	+1.64	−8.7
	V4r	✓		✓		✓		**+3.58**	**+1.55**	−3.1	**+7.30**	+0.56	−0.9
	V6	✓	✓	✓		✓	✓	**+4.60**	**+2.15**	−0.2	**+8.51**	**+6.21**	+2.2

Bold values highlight the significant improvement in performance

Table 9.6 A quantitative comparison of different fusion factors: the first letter in a group indicates the training set and the other refers to the test set. The letter K represents the KITTI dataset, and A denotes augmented data

	LAcc				mAcc			
	K-K	A-K	K-A	A-A	K-K	A-K	K-A	A-A
LiDAR	−1.27	−2.74	−4.64	−2.20	−0.63	−1.21	−2.65	−2.40
Dense	+2.98	+2.39	+1.35	+3.52	+1.48	+0.94	+1.19	+3.06
Early	+1.39	+0.46	+2.78	+1.10	+0.68	+0.17	+1.44	+2.15
Mid	−0.20	+0.60	+0.10	+1.46	−0.08	+0.39	+0.00	+0.50
Late	−1.42	−3.39	−8.98	−3.92	−0.68	−1.48	−4.29	−1.28
Road	+0.16	+3.49	+4.46	+3.79	+0.06	+1.58	+1.73	−0.88
adap	+1.04	+0.41	−1.46	+0.85	+0.55	+0.29	−0.20	+3.78

Actually, when we focus on the same row, we notice that early fusion outperforms midway the middle fusion when training on the KITTI, but the case will be inverse on the augmented dataset, and the late fusion performs even worse without sufficient capacity in fusion channel. In addition, the ratio $early/middle$ drops when the training set is changed from K to A or the test set changes from A to K; however, it increases when the order of the sets is reversed, from A-K to K-A. The contrast is also obvious if we normalize the value in the table by $\hat{x}_i = (x_i - x_{\min})/\Sigma(x_i - x_{\min})$. The main divergence in these train/test cases is the uncertainty, and the results verify our hypothesis about pre-fusion and post-fusion, as well as the impact of source. Our conclusion is applicable to situations in which a portion of sensors fail to record accurately or even break down, and we also present the result in the modality loss experiment.

Apart from these, we also learn the raw point clouds can be detrimental to the network, because the sparse signal generates poor SNR and reduces the channel capacity. In contrast, point cloud completion will compensate for this weakness, which is in reality the most advantageous aspect of fusion. Besides, multi-task learning (road) and adaptive fusion gain more capacity in almost all cases. Especially, the trend of road segmentation is close to the one of middle fusion as they serve similarly to increase the channel capacity. The adaptive fusion performs better when the training set and test set are similar, like K-K and A-A, although the result is more likely to be irregular. Possible reasons include the shortcomings of our simple design, competition with multi-task (they keep around 37% normalized contribution in all cases), or the overfitting on the training set, because it only optimizes the capacity allocation of the given channel based on the input data. It needs more experiments to acquire the convincing illustration.

Some testing examples are presented in Fig. 9.14. The line predictions in the last four rows are much more clearer than the others, because their predicted lines occupy more pixels on the boundaries, which is acceptable in usage and can be refined with additional post-processing.

It is unavoidable that error occurs to some of the sensors or data registration, both of which can result in the modality loss in the model. Consequently, we use zero

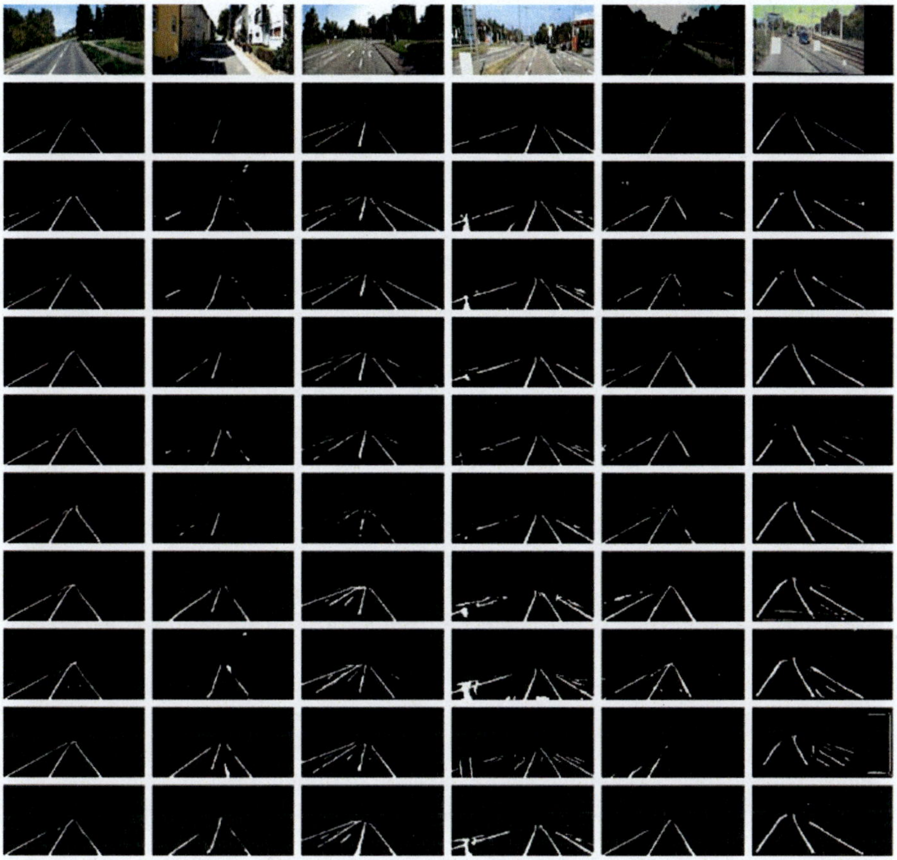

Fig. 9.14 Examples of the test result on the KITTI dataset: The three left columns are tested on the KITTI data and the rest are tested with the augmented data. The rows from top to bottom are input images, ground truth, and output from V1 to V5, V3r, V4r, V3r+, and V6

tensors as the lost data (other data is maintained) to replicate the testing situation and assess the robustness of models. The weight of single modality in the fused codes (features) can also be deduced according to Theorem 4. Moreover, the models are trained with augmented data and tested with KITTI.

As illustrated earlier, early fusion can be integrated into other tasks for an improved post-fusion, while middle fusion at later stage can fit the data more with deeper individual processes for different modalities. As shown in Table 9.7, V3/V3r surpass V4/V4r with only images, and a case never appears in the training set (the augmentation is mainly for images), while the latter models perform better with only point clouds. In general, in our experiment setting, middle fusion is more stable with modality loss. Besides, road segmentation as multi-task learning can extend the overall code length and capacity, thus enhancing error correction for both data.

Table 9.7 Experiments with single-modal data loss: The values in the last group (random) are the average of the two modality loss cases

	Image+points		Only image		Only points		Random	
Model	LAcc	mAcc	LAcc	mAcc	LAcc	mAcc	LAcc	mAcc
V3	85.24	91.89	56.74	77.76	20.29	60.15	38.52	68.98
V4	85.70	91.67	51.70	74.04	48.88	74.44	50.29	74.24
V3r	**89.04**	92.49	63.09	80.17	37.36	66.72	50.23	73.44
V4r	**88.87**	91.18	55.54	74.86	**67.88**	83.47	61.71	79.17
V6	**89.89**	92.69	**75.77**	85.89	50.34	74.51	**63.06**	80.21

Bold values highlight the significant improvement in performance

Besides, V6 performs better than V3/V3r, but utilizes fewer point clouds than V4r. Compared to V3/V3r, V6 not only adds the post-fusion stage by parallel coding after the first fusion, which helps to increase the SNR in the channel, but also extends the code length and capacity by adding external information of the point clouds in the middle stage and achieves better coding on the source. In other words, V6 acquires better post-fusion and distortion/redundancy balance than V3/V3r. Besides, the two-stage fusion forces the channel to allocate larger weight for point clouds than V3/V3r, resulting in better balanced results for both scenarios. However, for the different architecture V4r, the X-shaped fusion makes the transmission in point cloud pipeline to be influenced by the images, finally reducing the its capacity. In conclusion, V6 or other multi-stage fusion architectures essentially change the balance in capacity allocation. The result also shows that the global optimization in channels is not always accompanied by the local optimization.

Moreover, according to Theorem 4, the modality loss test also reflects the error correction capability of single-modal data and further implies the actual weight in the fusion. In particular, import modality would consume a considerable amount of space in the fused code and likely to necessitate a more extensive amount of code in the event that data is lost. Namely, the code performs worse without it. Based on the result in Table 9.7, V3/V4/V3r/V6 are better with only images, which means images contribute more in these lane line segmentation models. Then we can apply corresponding capacity allocation in the channel.

9.4 Summary

This chapter examines the robustness of single-modal and multimodal models under noisy data and analyzes and reveals the robustness of multimodal models on dirty data by comparing the accuracy decay of different models through experiments on the KITTI dataset. The experimental results show that the impact of noisy data on single-modal and multimodal models is complex and involves many factors like recognition target, recognition task, data modality combinations, noise type, etc. We also propose a novel multimodal fused 2D target detection model. It performs selective fusion of bounding boxes generated by multiple independent sub-models based on uncertainty estimation and is applicable to a variety of modal data. Experimental results show that our fusion model presents a high detection accuracy when a clean data modality is present under severe noise interference. In addition, the robustness of the fusion model may show different results for different fusion structures, such as feature fusion. The robustness problem of multimodal fusion models requires further in-depth study, and we will also focus on other tasks such as semantic segmentation and 3D target detection in our subsequent research to further investigate the phenomena observed in this chapter, while we also hope our experiments can provide some inspirations to other researchers.

We propose a novel camera-LiDAR fusion model for lane line segmentation. By leveraging the information from different sensors, the model can achieve the cutting-edge performance on the KITTI benchmark without pre-training or post-processing. Furthermore, we construct the multimodal network in terms of channels and utilize Shannon's theory to reveal the fusion mechanism. Based on the analysis, we consider it necessary to strike a balance among source, channel, and capacity. This chapter also provides practical methods to compare the contribution of different modalities, methods, and fusion stages, which will lead to the ideal fusion structure. Experiments have shown the benefits from information-driven fusion strategy and architecture. In the future, we will continue the work on quantification like uncertainty estimation and utilize more about joint coding model in the network. Besides, our work is supposed to be integrated well with other areas like detection and localization, which would be helpful for the future development of deep multimodal learning and autonomous driving.

References

1. Belghazi, M.I., Baratin, A., Rajeshwar, S., Ozair, S., Bengio, Y., Courville, A., Hjelm, D.: Mutual information neural estimation. In: International Conference on Machine Learning, pp. 531–540 (2018)
2. Bijelic, M., Muench, C., Ritter, W., Kalnishkan, Y., Dietmayer, K.C.J.: Robustness against unknown noise for raw data fusing neural networks. In: 2018 21st International Conference on Intelligent Transportation Systems (ITSC), pp. 2177–2184 (2018)
3. Bijelic, M., Gruber, T., Mannan, F., Kraus, F., Ritter, W., Dietmayer, K.C.J., Heide, F.: Seeing through fog without seeing fog: Deep multimodal sensor fusion in unseen adverse weather. In: 2020 IEEE/CVF Conference on Computer Vision and Pattern Recognition (CVPR), pp. 11679–11689 (2020)
4. Caesar, H., Bankiti, V., Lang, A.H., Vora, S., Liong, V.E., Xu, Q., Krishnan, A., Pan, Y., Baldan, G., Beijbom, O.: nuscenes: A multimodal dataset for autonomous driving. In: 2020 IEEE/CVF Conference on Computer Vision and Pattern Recognition (CVPR), pp. 11618–11628 (2020)
5. Chouikhi, N., Ammar, B., Hussain, A., Alimi, A.M.: Bi-level multi-objective evolution of a multi-layered echo-state network autoencoder for data representations. Neurocomputing **341**, 195–211 (2019). https://doi.org/10.1016/j.neucom.2019.03.012
6. Feng, D., Cao, Y., Rosenbaum, L., Timm, F., Dietmayer, K.: Leveraging uncertainties for deep multi-modal object detection in autonomous driving. In: 2020 IEEE Intelligent Vehicles Symposium (IV), pp. 877–884. IEEE, Piscataway (2020)

7. Feng, D., Haase-Schuetz, C., Rosenbaum, L., Hertlein, H., Duffhauss, F., Gläser, C., Wiesbeck, W., Dietmayer, K.C.J.: Deep multi-modal object detection and semantic segmentation for autonomous driving: datasets, methods, and challenges. IEEE Trans. Intell. Transport. Syst. **22**, 1341–1360 (2021)
8. Geiger, A., Lenz, P., Urtasun, R.: Are we ready for autonomous driving the kitti vision benchmark suite. In: 2012 IEEE Conference on Computer Vision and Pattern Recognition, pp. 3354–3361 (2012)
9. Gutmann, M., Hyvärinen, A.: Noise-contrastive estimation: A new estimation principle for unnormalized statistical models. In: Proceedings of the Thirteenth International Conference on Artificial Intelligence and Statistics, pp. 297–304 (2010)
10. He, K., Wang, Z., Fu, Y., Feng, R., Jiang, Y.G., Xue, X.: Adaptively weighted multi-task deep network for person attribute classification. In: Proceedings of the 25th ACM International Conference on Multimedia, pp. 1636–1644 (2017). https://doi.org/10.1145/3123266.3123424
11. He, Y., Zhu, C., Wang, J., Savvides, M., Zhang, X.: Bounding box regression with uncertainty for accurate object detection. In: Proceedings of the IEEE/CVF Conference on Computer Vision and Pattern Recognition, pp. 2888–2897 (2019)
12. Hinton, G., Vinyals, O., Dean, J.: Distilling the knowledge in a neural network (2015). Preprint arXiv:1503.02531
13. Hnewa, M., Radha, H.: Object detection under rainy conditions for autonomous vehicles: a review of state-of-the-art and emerging techniques. IEEE Signal Process. Mag. **38**, 53–67 (2021)
14. Kendall, A., Gal, Y.: What uncertainties do we need in bayesian deep learning for computer vision? In: NIPS (2017)
15. Khodabandeh, M., Vahdat, A., Ranjbar, M., Macready, W.G.: A robust learning approach to domain adaptive object detection. In: 2019 IEEE/CVF International Conference on Computer Vision (ICCV), pp. 480–490 (2019)
16. Kuleshov, V., Fenner, N., Ermon, S.: Accurate uncertainties for deep learning using calibrated regression (2018). ArXiv abs/1807.00263
17. Liang, M., Yang, B., Wang, S., Urtasun, R.: Deep continuous fusion for multi-sensor 3d object detection. In: ECCV (2018)
18. Liu, W., Anguelov, D., Erhan, D., Szegedy, C., Reed, S., Fu, C.Y., Berg, A.C.: Ssd: Single shot multibox detector. In: European Conference on Computer Vision, pp. 21–37. Springer, Berlin (2016)
19. Ma, W., Chen, J., Du, Q., Jia, W.: Pointdrop: Improving object detection from sparse point clouds via adversarial data augmentation. In: 2020 25th International Conference on Pattern Recognition (ICPR), pp. 10004–10009. IEEE, Piscataway (2021)
20. MacKay, D.J., Mac Kay, D.J.: Information Theory, Inference and Learning Algorithms. Cambridge University Press, Cambridge (2003). https://doi.org/10.1109/TIT.2004.834752
21. Michaelis, C., Mitzkus, B., Geirhos, R., Rusak, E., Bringmann, O., Ecker, A.S., Bethge, M., Brendel, W.: Benchmarking robustness in object detection: Autonomous driving when winter is coming (2019). ArXiv abs/1907.07484
22. Oh, S.I., Kang, H.B.: Object detection and classification by decision-level fusion for intelligent vehicle systems. Sensors **17**, 207 (2017)
23. Redmon, J., Farhadi, A.: Yolov3: An incremental improvement (2018). Preprint arXiv:1804.02767
24. Sun, P., Kretzschmar, H., Dotiwalla, X., Chouard, A., Patnaik, V., Tsui, P., Guo, J., Zhou, Y., Chai, Y., Caine, B., Vasudevan, V., Han, W., Ngiam, J., Zhao, H., Timofeev, A., Ettinger, S.M., Krivokon, M., Gao, A., Joshi, A., Zhang, Y., Shlens, J., Chen, Z., Anguelov, D.: Scalability in perception for autonomous driving: Waymo open dataset. In: 2020 IEEE/CVF Conference on Computer Vision and Pattern Recognition (CVPR), pp. 2443–2451 (2020)
25. Tian, J., Cheung, W., Glaser, N., Liu, Y.C., Kira, Z.: Uno: Uncertainty-aware noisy-or multimodal fusion for unanticipated input degradation. In: 2020 IEEE International Conference on Robotics and Automation (ICRA), pp. 5716–5723. IEEE, Piscataway (2020)

26. Tishby, N., Zaslavsky, N.: Deep learning and the information bottleneck principle. In: 2015 IEEE Information Theory Workshop (ITW), pp. 1–5 (2015). https://doi.org/10.1109/ITW.2015.7133169
27. Tishby, N., Pereira, F., Bialek, W.: The information bottleneck method. In: Proceedings of the 37th Allerton Conference on Communication, Control and Computation, vol. 49 (2001)
28. Xu, D., Anguelov, D., Jain, A.: Pointfusion: Deep sensor fusion for 3d bounding box estimation. In: 2018 IEEE/CVF Conference on Computer Vision and Pattern Recognition, pp. 244–253 (2018)
29. Yang, T., Li, Y., Ruichek, Y., Yan, Z.: Lanoising: A data-driven approach for 903nm tof lidar performance modeling under fog. In: 2020 IEEE/RSJ International Conference on Intelligent Robots and Systems (IROS), pp. 10084–10091 (2020)
30. Zhuang, F., Qi, Z., Duan, K., Xi, D., Zhu, Y., Zhu, H., Xiong, H., He, Q.: A comprehensive survey on transfer learning. Proc. IEEE **109**, 43–76 (2021)
31. Zou, Z., Li, Y.: Efficient urban-scale point clouds segmentation with bev projection (2021). ArXiv abs/2109.09074
32. Zou, Z., Zhang, X., Liu, H., Li, Z., Hussain, A., Li, J.: A novel multimodal fusion network based on a joint coding model for lane line segmentation. Informat. Fusion **80**, 167–178 (2022)

Chapter 10
Conclusions

With the development of vehicle digitization and intelligence, the multimodal fusion of different types of sensor data is an inevitable trend to realize all-weather and all-scenario autonomous driving perception. Hence, this book proposes innovative solutions based on multi-sensor fusion for various tasks in autonomous driving perception. With multimodal fusion technology as the core, the following autonomous driving perception problems are systematically solved.

1. As the basis of multimodal fusion, Chap. 3 designs an automated online calibration technology based on a monocular camera and 3D LiDAR for the perception system of intelligent driving vehicles. It effectively overcomes the influence of the sensor space position change caused by the load change or vibration on the system. The proposed technology realizes accurate automatic calibration between sensors during vehicle driving, which provides a foundation for subsequent work.
2. For object detection and scene segmentation tasks in autonomous driving, various fused perception algorithms are proposed in Chaps. 4, 5. A fusion mechanism with high fusion accuracy, low network complexity, and low inference delay is explored, which significantly improves the performance of existing lane/road segmentation models. A 3D target detection technology based on fusion point cloud enhancement and 4D millimeter-wave RaDAR is proposed, effectively improving recognition accuracy and reliability.
3. Precise positioning technology is also an essential part of autonomous driving perception. To solve the problem of satellite navigation and positioning system failure in complex dynamic scenarios, Chap. 6 proposes an autonomous fusion positioning method with high precision and versatility. An autonomous localization method combined with the semantic map generation method is designed, and a fusion localization technology based on global information and feature SLAM is proposed. The above technology can quickly eliminate the accumulated errors in the existing positioning methods and improve the positioning accuracy.

X. Zhang et al., *Multi-sensor Fusion for Autonomous Driving*,
https://doi.org/10.1007/978-981-99-3280-1_10

4. In order to further improve the perception system of intelligent networked vehicles, Chaps. 7–9 purpose a series of advanced content from the perspective of dataset, vehicle-road collaboration, and information quality. In the subsequent research stage, key technical details will be optimized to improve the accuracy and application scope of the algorithm and better serve the development of autonomous vehicle technology.

Using the developed multimodal fusion method, a series of autonomous driving perception problems including object detection, segmentation, and localization can be addressed. However, the problem of limited vehicle perception range in complex mixed traffic environments has gradually become prominent. Autonomous driving perception is evolving from "single-vehicle intelligence" to "vehicle-road coordination." As an extension of the algorithm proposed in this book, using the redundant and complementary advantages of multi-source heterogeneous sensors, the vehicle-road collaborative perception of autonomous vehicles in complex environments can be realized in the future.